STUDY GUIDE
STUDENT SOLUTIONS MANUAL

College Physics

Second Edition

STUDY GUIDE
STUDENT SOLUTIONS MANUAL

John Kinard/Jerry D. Wilson

College Physics

Second Edition

Jerry D. Wilson
Lander University

Prentice Hall, Englewood Cliffs, New Jersey 07632

Editorial/production supervision: ***Amy K. Jolin***
Acquisitions editor: ***Paul Banks***
Supplements acquisitions editor: ***Mary Hornby***
Production coordinator: ***Trudy Pisciotti***

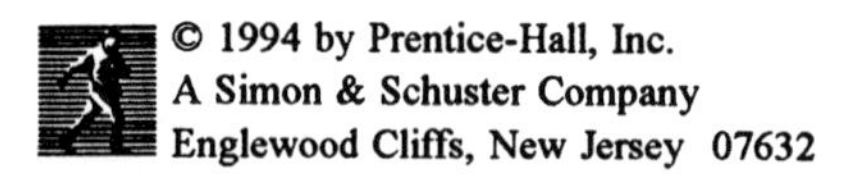

A Simon & Schuster Company
Englewood Cliffs, New Jersey 07632

Printed in the United States of America

10 9 8 7 6

ISBN 0-13-147612-2

PRENTICE-HALL INTERNATIONAL (UK) LIMITED, LONDON
PRENTICE-HALL OF AUSTRALIA PTY. LIMITED, SYDNEY
PRENTICE-HALL CANADA INC. TORONTO
PRENTICE-HALL HISPANOAMERICANA, S.A., MEXICO
PRENTICE-HALL OF INDIA PRIVATE LIMITED, NEW DELHI
PRENTICE-HALL OF JAPAN, INC., TOKYO
SIMON & SCHUSTER ASIA PTE. LTD., SINGAPORE
EDITORA PRENTICE-HALL DO BRASIL, LTDA., RIO DE JANEIRO

Preface

This **Study Guide and Student Solutions Manual** for **College Physics**, second edition, was prepared to help students gain a greater understanding of the principles of their introductory physics course. Most of us learn by summary and example, and this Manual has been organized along these lines. For each chapter you will find :

*Chapter Objectives

*Chapter Summary

*Important Terms and Relationships

*Additional Solved Problems

*Solutions to Selected Exercises and Paired Exercises

*Sample Quiz (with answers)

Let's look at each of these features briefly.

* The Chapter Objectives section states the learning goals for the chapter. The objectives tell you what you should know upon the completion of chapter study.

* The Chapter Summary section outlines the important chapter concepts and provides a brief overview of the major chapter contents. This review allows you to check the thoroughness of your study.

* The Important Terms and Relationships section lists the important (bold face) terms and equations -- by chapter section. These are similar to the Important Concepts and Relationships at the end of the text chapters. The purpose is for self-review. You should be able to clearly define and explain each important term. (If not, a review of the Chapter Summary section should help.) Also, you should be able to identify each symbol in an equation and explain what relationship the equation describes.

* The Additional Solved Problems section provides two to three example problems for each chapter section, thus giving twelve to fifteen solved examples per chapter. Seeing how example problems are solved is a great learning benefit. You should find this section very helpful.

* The Solutions to Selected Exercises and Paired Exercises section provides the worked out solutions of annotated (blue dot) end-of-chapter text exercises. The paired exercises are similar in nature. You should try to work out the first even-

numbered paired text exercise independently and then check work with the solution in this Manual. After working the odd-numbered exercise of a pair, you can check your answer in the Answers to Odd-Numbered Exercises at the back of the text.

* Finally, the Sample Quiz , which consists of multiple choice questions and problems, allows you to self-check your knowledge of a chapter. The answers to the quizzes are given at the back of the Manual.

As you can see, the Manual provides a thorough review for each chapter. The conscientious student can make good use of the various sections to assist in understanding and mastering the course contents. We hope you find it so.

John L. Kinard

Jerry D. Wilson

Contents

CHAPTER 1 Units and Problem Solving

Chapter Objectives

Upon completion of your study in Chapter 1 - **Units and Problem Solving** students should be able to :

1. describe the SI and explain its advantages,
2. distinguish between a base unit and a derived unit.
3. state the basic SI units used to measure length, time, mass, and capacity.
4. interpret and apply metric prefixes.
5. check for the correctness of an equation through the use of dimensional analysis.
6. convert units within the SI, within the British system, and between the two measurement systems.
7. explain the reason for using significant digits.
8. state the correct siginifiant digits in a number and write digits in scientific (powers of 10) notation.
9. add, subtract, multiply, and divide to the correct number of significant digits.
10. state the equation for mass density and solve problems on mass density.
11. apply the process for problem solving to the solution of exercises.

Chapter Summary

Objects and phenomena are measured and described using **standard units**, a group of which makes up a **system of units.**

* The metric **SI** has seven base units (Table 1.1). The base units for length, mass, and time are the **meter** (m), the **kilogram** (kg), and the **second** (s), respectively.

* The metric system is a decimal or base-10 system, which is convenient in changing measurements from one size unit to another. Metric multiples are designated by prefixes, the most common of which are kilo-(1000), centi-(1/100), and milli-(1/1000). A unit of volume or capacity is the **liter** (L), and 1 L = 1000 mL = 1000 cm^3 (cubic centimeters).

*Fundamental base quantities, such as length, mass, and time are called dimensions. These are commonly expressed by bracketed symbols [L], [M], and [T], respectively. **Dimensional analysis** is a procedure by which the dimensional correctness of an equation may be checked. Both sides of an equation must not only be equal in numerical magnitude, but also in dimension, and dimensions can be treated like algebraic quantities. Units, instead of symbols, may be used in unit (dimensional) analysis.

*A quantity may be expressed in other units through the use of **conversion factors**, i.e. (1in. / 2.54 cm) or (1cm / 0.395 in.). The appropriate form of a conversion factor for mathematical operation is easily determined by unit analysis.

* The number of **significant figures** of a quantity is the number of reliably known digits in the quantity, In general, the final result of a multiplication and/or division operation should have the same number of significant figures as the quantity with the least number of significant figures used in the calculation. The final result in addition (or subtraction) should have the same number of decimal places as the quantity with the least number of decimal places used in the calculation. The proper number of figures or digits is obtained by rounding off a result.

* Suggested problem solving procedure :
  1. List data and what is to be found.
  2. Draw a diagram.
  3. Determine which principle(s) and equation(s) are relevant and how to use them.
  4. Simplify equations algebraically.
  5. Check units and make necessary conversion of units.
  6. Perform calculations, observing significant figures.
  7. Consider whether the result is reasonable, i.e., whether it has an appropriate magnitude.

Important Terms and Relationships

1.1 Why and How We Measure

standard unit
systems of units

1.2 SI Units for Length, Mass, and Time

SI (International System of Units)
SI base units
SI derived units
meter (m)
kilogram (kg)
second (s)

1.3 More about the Metric System

mks system
cgs system
fps system
liter (L)

1.4 Dimensional Analysis

dimensional analysis
unit analysis
density : $\rho = m / V$

1.5 Unit Conversions

unit conversions
conversion factor

1.6 Significant Figures

exact numbers
significant figures
measured number
rules for using signifiant figures

1.7 Problem Solving

problem solving
seven-step process for solving problems

Additional Solved Problems

1.4 Dimensional Analysis

Example 1 Show that the equation $R = v_o^2 \sin 2\theta / g$ is dimensionally correct where v_o is the velocity and g is the acceleration.

Solution Using the appropriate dimensions, [L / T] for v_o and $[L/T^2]$ for g.

$[L] = [L / T]^2 / [L/T^2]$
$[L] = [L^2 / T]^2 \times [T^2] / [L]$
$[L] = [L]$

Example 2 The area for a circle is $A = \pi r^2$. Use SI unit analysis to find the units for area.

Solution The standard SI length unit is meter (m) .

$A = \pi\ (m)^2$

Area has the units of $[L]^2$ and π is dimensionless.

1.5 Unit Conversions

Example 1 Find the length of a 100 yd. dash in meters.

Solution Using the appropriate conversion factors :

(100 yd) (3 ft / yd) (1 m / 3.28 ft) = 91.5 m

Example 2 The speed of a body is measured to be 100 m/s. Find the corresponding speed in mi/h.

Solution One method is to convert units of length and time separately :

(100 m/s) (1 mi / 1609 m) (3600 s / h) = 224 mi/h

A second solution is using a direct conversion factor :

(100 m/s) (2.24 mi/h / m/s) = 224 mi/h

Example 3 The dimensions a room are 20 ft by 16 ft by 9.0 ft high.
A. Find the volume of the room in ft^3.
B. Find the volume of the room in m^3.
C. Find the volume of the room in liters.

Solution First find the volume for a rectangular solid.

V = L x w x h
V = (20 ft) x (16 ft) x (9.0 ft)
V = 2.9×10^3 ft^3

Now convert the units from ft^3 to m^3 using a conversion factor.

V = (2.9×10^3 ft^3) (0.0283 m^3 / ft^3)
V = 82 m^3
Finally convert the units from m^3 to L

V = (82 m^3) (10^3 L / m^3)
V = 8.2×10^4 L

Example 4 The area for a triangle is 50 cm^2. The length of the altitude (a) of the triangle is 1.5 m. What is the measure of the base (b) of the triangle in cm ?

Solution The area for a triangle = (1/2) ab. Since the answer is to be expressed in cm, convert the altitude to cm.

(1.5 m) (100 cm / m) = 150 cm
A = (1/2) ab

$50\ cm^2 = (1/2)\ (150\ cm)\ (\ b\ )$
$50\ cm^2 = 75\ cm\ (\ b\ )$
$h = 0.67\ cm$

1.6 Significant Figures

Example 1 How many significant figures are there in the following numbers ? Place your answers in the appropriate space.

A. 56.2 ________
B. 5005 ________
C. 0.0060 ________
D. 2.0×10^6 ________
E. 2.50×10^{-2} ________

Solution

A. three - non-zero digits are significant.
B. four - zeros interior to a number are significant.
C. two - zeros at the beginning of a number are not significant.
D. two - zeros at the end of a number after the decimal point are significant.
E. three - same reason as in D

Example 2 Write the following digits in scientific notation to three significant digits.

A. 0.004567 ____________
B. 456,400 ____________
C. 186,000 ____________
D. 0.000000008999 ____________
E. 56.865 ____________

Solution The final result of a multiplication or division operation should have the same number of significant figures as the quantity with the least number of significant figures that was used in the calculation.

A. 4.57×10^{-3}
B. 4.56×10^5
C. 1.86×10^5

D. 9.00×10^{-9}

E. 5.69×10^{1}

Example 3 Perform the indicated operations observing the rules for significant digits.

A. 23.1 + 45 + 0.68 + 100 = ___________

B. 157 - 5.689 + 2 = ___________

C. 23.5 + 0.567 + 0.85 = ___________

D. $4.69 \times 10^{-6} - 2.5 \times 10^{-5}$ = ___________

E. $8.9 \times 10^{4} + 2.5 \times 10^{5}$ = ___________

Solution The final result of the addition should have the same number of decimal places as the quantity with the least number of decimal places.

A. 23 + 45 + 1 + 100 = 169

B. 157 - 6 + 2 = 153

C. 23.5 + 0.6 + 0.9 = 25.0

D. re-write 2.5×10^{-5} to 25×10^{-6}

$(4.69 - 25) \times 10^{-6} = (5 - 25) \times 10^{-6} = -20 \times 10^{-6} = -2.0 \times 10^{-5}$

E. re-write 2.5×10^{5} to 25×10^{4}

$(8.9 + 25) \times 10^{4} = 34 \times 10^{4} = 3.4 \times 10^{5}$

Example 4 Perform the following operations following the rules for significant digits.

A. 0.568 x 3.4 = ___________

B. 13.90 ÷ 0.580 = ___________

C. $4.8 \times 10^{5} \div 4.0 \times 10^{-3}$ = ___________

D. $(3.2 \times 10^{8})(4.0 \times 10^{4})$ = ___________

Solution Multiply or divide then round to the correct number of significant figures.

A. 1.9312 to two significant digits = 1.9

B. 23.9655 to three significant digits = 24.0

C. first divide 4.8 by 4.0 = 1.2 ; then 5 - (-3) = 8 ; 1.2×10^{8}

D. first multiply 3.2 by 4.0 = 12.8 = 13 ; add exponents 12 ; $13 \times 10^{12} = 1.3 \times 10^{13}$

1.7 Problem Solving

Example 1 The density of air is 1.29 kg/m^3. Find the mass of the air in the room having the following dimensions : 13 ft x 16 ft x 9.0 ft.

given : the dimensions of the room : 13 ft x 16 ft x 9.0 ft
the density of air : 1.29 kg / m^3

First find the volume of the room. Be aware of the different units in the problem. The problem may be worked several ways correctly.

$L = 13\ ft\ (1\ m / 3.28\ ft) = 4.0\ m$
$W = 16\ ft\ (1\ m / 3.28\ ft) = 4.9\ m$
$h = 9.0\ ft\ (1\ m / 3.28\ ft) = 2.7\ m$
$V = L \times W \times h$
$V = (4.0\ m)(4.9\ m)(2.7\ m)$
$V = 53\ m^3$

Next find the mass of the air.

$\rho = m / V$
$m = \rho V$
$m = (1.29\ kg / m^3)(53\ m^3)$
$m = 68\ kg$

Example 2 A 5.0 gallon bucket is filled to the brim with water. If the unfilled bucket has a mass of 5.0 kg, find the total mass of the bucket in kg. The density of water is 1.0×10^3 kg/m^3.

Given : mass of the bucket : $m_b = 5.0$ kg
mass of the water is unknown, but the volume of the water ($V_w = 5.0$ gal) and the density of water are both known.

First find the volume of the water in m^3.
$V_w = 5.0\ gal\ (3.785\ L / gal)(1.0\ m^3 / 1000\ L) = 1.9 \times 10^{-2}\ m^3$
The total mass $M = m_w$ (mass of water) + m_b (mass of the bucket)
$\rho_w = m_w / V_w$

$m_w = \rho_w V_w$

$m_w = (1.0 \times 10^3\ kg/m^3)\ (1.9 \times 10^{-2}\ ft^3) = 19\ kg$

The total mass to be lifted $M = m_w + w_b = 19\ kg + 5\ kg = 24\ kg$

Example 3 A car can travel a distance of 200 miles in three hours 40 minutes. If the car travels at the same average rate, how far can the car travel in 5.0 hours.

given : $d_1 = 200$ miles ; $t_1 = 3.0\ h + 40\ min = 3.7\ h$

$d_2 = x$; $t_2 = 5.0\ h$

since the rate of travel is the same a proportion will work nicely.

$d_1 / d_2 = t_1 / t_2$

$200\ mi / d_2 = 3.7\ h / 5.0\ h$

$(3.7)\ d_2 = 1.0 \times 10^3$

$d_2 = 2.7 \times 10^2\ mi$

Solutions to the paired problems and other problems

6. (d) magnitude of units, units, and dimensions should all be the same for both sides of an equation.

12. Using the appropriate dimensions, $[m]^2$ for A and [m] for r.

$A = 4 \pi r^2$

$[m^2] = 4 \pi [m]^2$

$[m^2] = [m^2]$

16. Using the appropriate units : [m] for x and [s] for t.

$x = gt^2 / 2$

solve for g in the equation :

$x / t^2 = g$

m / s^2 are the units for g

22. Using the appropriate units : [m] for x , [kg] for m, and [m/s^2] for g
$mg = kx$
solve for k : $mg / x = k$ or $k = mg / x$
$k = (kg)(m/s^2) / m$
$k = kg / s^2$

34. first 100 m : (100 m) (1.0×10^2 cm / m) = 1.0×10^4 cm

now convert 100 yd to cm
(100 yd) (3 ft / yd) (30.48 cm / ft) = 9.1×10^3 cm
the difference is 1.0×10^4 cm - 0.91×10^4 cm = 0.1×10^4 cm = 1.0×10^3 cm

39. first miles/gal : 750 miles / 30.0 gal = 25.0 mi/gal
km / L : (750 mi)(1.609 km/mi) = 1.21×10^3 km
(30.0 gal)(3.785 L / gal) = 114 L
(1.21×10^3 km) / (114 L) = 10.6 km/L
m / mL (1.21×10^3 km) (1000 m/km) = 1.21×10^6 m
(114 L)(1000 mL / L) = 1.14×10^5 mL
(1.21×10^6 m) / (1.14×10^5 mL) = 10.6 m / mL

44. (a) convert the dimensions from ft to m :
(300 ft) (0.3048 m / ft) = 91.4 m
(160 ft) (0.3048 m / ft) = 48.8 m
now find the dimensions in cm :
(91.4 m) (100 cm / m) = 9.14×10^3 cm
(48.8 m) (100 cm / m) = 4.88×10^3 cm
The area can be found by finding the area for a rectangle.
$A = (9.14 \times 10^3 \text{ cm})(4.88 \times 10^3 \text{ cm}) = 4.46 \times 10^7 \text{ cm}^2$

53. (a) 4 significant figures since zeros are between two non-zero digits
(b) 3 significant figures since zero is between two non-zero digits
(c) 5 since there all are non-zero
(d) 2 since zeros to the left of digit 1 are to show place value

58. $A = (1/2)\,ab$
$A = (1/2)(10.5 \text{ cm})(8.7 \text{ cm})$
$A = 45.675 \text{ cm}^2$ to two significant digits is 46 cm^2

68. given : mass $m = 6.0 \times 10^{24}$ kg
volume $V = 1.1 \times 10^{21}\ m^3$
find : density ρ
$\rho = m / V$
$\rho = (6.0 \times 10^{24}\ kg) / (1.1 \times 10^{21}\ m^3)$
$\rho = 5.5 \times 10^3\ kg / m^3$

74. cost : C = \$6.50
need to find the number of kg in three pounds

(3 lb) (1 kg / 2.2 lb) = 1.36 kg

\$6.50 / 1.36 kg = \$4.78 / kg

83. area for a sphere $A = 4\pi\ r^2$
(a) $A = 4\pi\ (12)^2 = 1.8 \times 10^3\ cm^2$
(b) $(1.8 \times 10^3\ cm^2)(10^{-4}\ m^3 / cm^3) = 0.18\ m^3$
(c) first the volume must be found : $V = (4/3)\ \pi r^3$;
$V = (4/3)\pi(0.12)^3 = 7.24 \times 10^{-3}\ m^3$
$\rho = m / V$; $\rho = (4.0\ kg) / (7.24 \times 10^{-3}\ m^3) = 5.6 \times 10^2\ kg / m^3$

Sample Quiz

Completion

1. In the SI , the basic unit for measuring time is the ___________, the basic unit for length is the _________________, and the basic unit for measuring mass is the _____________________.

2. 3.00×10^5 m + 2.5×10^4 m = ______________________

3. If **c** has units of m/s^3 and **a** has the units of s, the unit of the term which has the product **ac** is ____________________.

Multiple Choice. Choose the correct answer.

_____4. When (3.5×10^4) is multiplied by (4.00×10^2), the product is which of the following expressed to the correct number of siginificant digits.

A. 1.75×10^2
B. 1.4×10^6
C. 14×10^7
D. 1.4×10^7

_____5. Which of the following is not equivalent to 2.50 miles ?

A. 1.32×10^4 ft
B. 1.6×10^5 in
C. 4.03×10^3 km
D. 4.03×10^5 cm

____6. The area for a room is 15 m^2. How many cm^2 are there in the room ?

A. 0.15 cm^2
B. 1.5×10^{-3} m^2
C. 1.5×10^3 cm^2
D. 1.5×10^5 cm^2

Problems

7. Using unit analysis, show that the equation $\mathbf{y = (v_o \sin\theta)^2 / 2g}$ is dimensionally correct. y is length, v_o is velocity and g is acceleration.

8. A car makes a journey of 200 miles on 12 gallons of gas. The time for the car to travel this distance is 4.0 hours.
   A. Find rate at which gas is consumed during the trip.
   B. Find the speed at which the car travels.

9. In problem 8, how many liters of gasoline were used and how many kilometers were traveled during the trip?

10. The density of water is 1.0×10^3 kg/m^3. Find the mass of water needed fill
    A. a 3.0 L soda pop container.
    B. a rectangular box having the dimensions of 2.5 m x 3.0 m x 1.5 m.

CHAPTER 2 Kinematics

Chapter Objectives

Upon completion of the study on the unit on kinematics, students should be able to :

1. define the terms mechanics, kinematics, and dynamics.
2. distinguish between distance and displacement.
3. distinguish between and give examples of vectors and scalars.
4. define average velocity, average speed, and instaneous speed and velocity.
5. define acceleration and distinguish between zero acceleration, constant acceleration, and increasing or decreasing acceleration.
6. define free fall.
7. solve problems on constant speed and velocity.
8. write expressions for velocity and position as functions of time for objects moving with zero acceleration and with uniform acceleration.
9. apply constant acceleration equations to solve problems involving one-dimensional motion.
10. sketch graphs plotting position versus time, velocity versus time, and acceleration versus time for uniform velocity, uniform acceleration, and increasing or decreasing acceleration.
11. interpret graphs of position versus time, velocity versus time, and acceleration versus time so that they can find areas and slopes to give numerical analysis, and sketch other graphs quantitatively and qualitatively using a given graph.

Chapter Summary

* Motion is the changing of position. The length traveled in changing position may be expressed in terms of **distance**, the actual path length between two points. Distance is a scalar quantity (numerical magnitude only). The directed straight-line distance between two points is referred to a **displacement**. Displacement is a vector quantity, and has both magnitude and direction.

*In describing motion, rate of change of position may be expressed in several ways: **Average speed** is the distance traveled divided by the time to travel that distance. **Instantaneous speed** is the speed at a particular instant of time (Δt is close to zero). **Average velocity** is the displacement divided by the travel time. **Instantaneous velocity** is the velocity at a particular instant of time (Δt is close to 0). The SI units for measuring speed and velocity are m/s. Speed is a scalar quantity and velocity is a vector.

*Acceleration is the rate of change of velocity with time. **Average acceleration** is the rate of change of velocity divided by the time to make the change. **Instantaneous acceleration** is the acceleration at a particular instant of time (Δt is close to zero). The units for measuring acceleration are m/s^2. An acceleration may result from a change in magnitude and / or the direction of velocity (a vector).

*For linear motion, when the velocity and acceleration are in the same direction, and they have the same directional signs, the velocity increases. When the velocity and acceleration are in opposite directions, different signs, the velocity decreases.

*In general, linear motion with constant acceleration may be described by three general equations and two algebraic combinations of these equations that provide calculation convenience. These are listed under important formulas in your text.

*Objects in motion solely under the influence of gravity are said to be in **free fall**. Near the surface of the Earth, the **acceleration due to gravity (g)** has a relatively constant value of 9.80 m/s^2.

*When working with kinematic equations, you must take into account the direction of the vector quantities. For free fall, it is common to take upward as the positive direction and downward as the negative direction. The acceleration due to gravity is then always - g, which can be explicitly expressed in the kinematic equations.

Important Terms and Relationships

Introduction

mechanics
kinematics
dynamics

2.1 A Change in Position

motion
scalar quantity
vector quantity
distance
displacement

2.2 Speed and Velocity

speed :
- average
- instantaneous
- $s = \Delta d / \Delta t$

velocity
- average
- instantaneous
- $x = vt$

2.3 Acceleration

acceleration :
- average
- instantaneous

acceleration $a = \Delta v / \Delta t$

constant or uniform acceleration
- $v_{av} = (v + v_o) / 2$
- $v = v_o + at$
- $x = v_o t + (1/2)at^2$
- $v^2 = v_o^2 + 2ax$

2.4 Kinematic Equations (emphasis on problem solving)

2.5 Free Fall

free fall

acceleration due to gravity : $9.8\ m/s^2$

$y = v_o t - (1/2)gt^2$

$v = v_o - gt$

$v^2 = v_o^2 - 2gy$

Summary on graphs

Slope of x vs. t gives velocity
Slope of v vs. t gives acceleration
area of v vs. t gives displacement
area of a vs. t gives change in velocity

Additional Solved Problems

2.1 Distance - Displacement

A person walks 1.00 km east and then walks 0.75 km west.

A. What total distance has the person walked ?

B. What is the displacement of the walker relative to the initial position.

A. Distance is a scalar - therefore direction does not need to be considered.

$x = x_1 + x_2$; $x = 1.00 \text{ km} + 0.75 \text{ km} = 1.75 \text{ km}$

B. Displacement is a vector, therefore direction must be taken into consideration.

$\Delta x = x_1 - x_2$

$\Delta x = 1.00 \text{ km} - 0.75 \text{ km} = 0.25 \text{ km}$; the net displacement relative to the initial position is east.

$\Delta x = 0.25$ km east

2.2 Speed and Velocity

Example 1 A jogger can jog at the average rate of 10 km/h. How many minutes will it take the jogger to travel a distance of 6.0 km?

given : $s = 10$ km/h ; $d = 6.0$ km ; find t in h (time)

$s = \Delta d / \Delta t$

$10 \text{ km/h} = 6.0 \text{ km} / t$; $t = 0.60 \text{ h}$; $(0.60 \text{ h})(60 \text{ min} / \text{h}) = 36 \text{ min}$

Example 2 A motorist drives 80.0 km/h east for one hour and then drives 90.0 km/h for 60.0 km west.

A. What is the average speed for the motorist for the entire trip ?

B. What is the average velocity for the motorist for the entire trip ?

given : $v_1 = 80.0$ km/h east ; $t_1 = 1.00$ h

$v_2 = 90.0$ km/h west ; $x_2 = 60.0$ km west

first find x_1 $\quad x_1 = v_1 t_1$; $x_1 = (80.0 \text{ km/h})(1.00 \text{ h}) = 80.0$ km east

second find t_2 $\quad x_2 = v_2 t_2$; $60.0 \text{ km} = (90.0 \text{ km/h})\, t_2$; $t_2 = 0.67 \text{ h}$

A. $s = \Delta d / \Delta t$; $s = s_1 + s_2 = 80$ km + 60 km= 140 km ;

$t = t_1 + t_2$; $t = 1.0$ h + 0.67 k = 1.67 h

$s = 140$ km / 1.67 h = 83.4 km/h

B. velocity is a vector therefore the displacement should be found first.
$\Delta x = 20.0$ km east (since the displacement are in opposite directions)
$x = vt$; 20.0 km = v (1.67 h) ; v = 12.0 km/h east

2.3 Acceleration

Example 1 A car can accelerate at the rate of 4.0 m/s^2. How fast will the car be traveling after the following times (assuming the car starts from rest) ?
A. 1.0 s ?
B. 2.0 s ?
C. 3.0 s ?
D. 10 s ?

given : $a = 4.0$ m/s^2 (that means v increases 4.0 m/s each s) ; $v_o = 0$

$a = \Delta v / \Delta t$

A. 4.0 m/s^2 = Δv / 1.0 s ; Δv = 4.0 m/s

B. since a and Δt are same between 1.0 and 2.0 as they were from 0 to 1.0, Δv = 4.0 m/s ; after 2.0 s the car is traveling at 8.0 m/s

C. after 3.0 s the car's velocity has increased by 4.0 m/s from the previous time interval ; thus the velocity of the car is 12 m/s

D. since $a = \Delta v / \Delta t$ take Δv and Δt measured from rest (t = 0 and v = 0)
4.0 m/s^2 = Δv / 10 s ; Δv = 40 m/s ; therefore the car is traveling at 40 m/s

Example 2 An object is traveling with a velocity of 40 m/s in the +x direction when it experiences an acceleration of 2.0 m/s^2 in the -x direction.
A. How long will the object travel before it stops ?
B. How far does the object travel from its initial point when it stops ?
C. What is the velocity and the position of the object after 8.0 s ?
D. What is the velocity and the position of the object after 30 s ?
E. How long will it take the object to return to its initial position ?

given : for all parts v_o = 40 m/s and a = -2.0 m/s^2

A. when the object stops v = 0
$v = v_o - at$

0 = 40 m/s - (2.0 m/s^2) t ; t = 20 s

B. $x = v_o t + (1/2) at^2$

x = (40 m/s)(20 s) + (1/2)(-2.0 m/s^2)(20 s)^2
x = 800 m - 400 m ; x = 400 m in the positive x direction

C. take expressions for parts A and B and simply substitute 8.0 s for time.

v=40 m/s - (2.0 m/s^2)(8.0 s) x=(40 m/s)(8.0 s)+(1/2)(-2.0 m/s^2)(8.0 s)^2
v = 40 m/s - 16 m/s x = 320 m - 64 m

v = 24 m/s in the (+) x direction x = 2.6 x 10^2 m (+) x direction

D. the only difference from (C) is time.

v = 40 m/s - (2.0 m/s^2)(30 s) x=(40 m/s)(30 s)+(1/2)(-2.0 m/s^2)(30 s)^2

v = -20 m/s x = 300 m
Note that v is (-) and x is (+). What does this mean ? The object has reached its maximum point and is now on a return trip.

E. x = 0 ; 0 = (40) t + (1/2)(-2.0) t^2
t = 40 s

Example 3 A car initially at rest accelerates at 2.00 m/s^2 for 5.00 s, then moves with a constant velocity for 5.00 s and then decelerates at the rate of 2.0 m/s^2 for an additional 5.00 s.

A. What is the distance the car travels during the first 5.00 s ?
B. What is the distance the car travels during the second 5.00 s ?
C. What is the distance the car travels during the third 5.00 s ?
D. What is the total displacement of the car during this 15.0 s ?

given : v_o = 0 ; a = 2.00 m/s^2 ; t = 5.00 s ; find x ; given t

A. $x = v_o t + (1/2)at^2$

x = 0 + (1/2)(2.00 m/s^2)(5.00 s)^2 = 25.0 m

B. here the car is traveling with a constant velocity
x = vt (one problem what is v ?)
back to the acceleration interval : $v = v_o + at$

v = 0 + (2.00 m/s^2)(5.00 s) = 10.0 m/s
x = (10.0 m/s) (5.00 s) = 50.0 m

C. $x = v_ot + (1/2)at^2$

$x = (10.0 \text{ m/s})(5.00 \text{ s}) + (1/2)(-2.00 \text{ m/s}^2)(5.00 \text{ s})^2$ (note $v_o \neq 0$)

$x = 25$ m (note x is positive)

D. $x = x_1 + x_2 + x_3$

$x = 25 \text{ m} + 50 \text{ m} + 25 \text{ m} = 100 \text{ m}$

2.4 Kinematic Equations

Example 1 An electron starts from rest and is accelerated to a speed of 1.0×10^6 m/s through a distance of 20 cm. Find the acceleration of the electron.

Solutions given : $v_o = 0$; $v = 1.0 \times 10^6$ m/s $x = 20 \text{ cm} = 0.20 \text{ m}$

$v^2 = v_o{}^2 + 2ax$

$(1.0 \times 10^6 \text{ m/s})^2 = 0^2 + 2\,(a)(0.20 \text{ m})$

$a = 2.5 \times 10^{12} \text{ m/s}^2$

Example 2 A ball starts from rest rolls down an inclined plane with an acceleration of 2.0 m/s^2.

A. Find the distance the ball travels during each of the first three seconds.

B. Find the speed of the ball after the first, second, and third seconds.

given : $v_o = 0$; $a = 2.0 \text{ m/s}^2$

A. $x = v_ot + (1/2)at^2$

during the first second $x = (1/2)(2.0 \text{ m/s}^2)(1.0 \text{ s})^2 = 1.0 \text{ m}$

distance traveled from rest after 2.0 s : $x = (1/2)(2.0 \text{ m/s}^2)(2.0 \text{ s})^2$

$x = 4.0$ m

the distance traveled during the second second is 3.0 m

distance traveled from rest after 3.0 s : $x = (1/2)(2.0 \text{ m/s}^2)(3.0 \text{ s})^3$

$x = 9.0$ m

the distance traveled during the third second is 5.0 m

B. $v = v_o + at$

after the first second : $v_1 = 0 + (2.0 \text{ m/s})(1.0 \text{ s}) = 2.0 \text{ m/s}$

after two seconds : $v_2 = 0 + (2.0 \text{ m/s})(2.0 \text{ s}) = 4.0 \text{ m/s}$

after three seconds : $v_3 = 0 + (2.0 \text{ m/s})(3.0 \text{ s}) = 6.0 \text{ m/s}$

2.5 Free Fall

Example 1 A ball is thrown upward with an initial upward velocity of 29.4 m/s. It is later caught at the same height is was thrown.
A. How high will the ball rise ?
B. What is the velocity of the ball at the peak point ?
C. What is the acceleration of the ball at the peak point ?
D. How long does it take the mass to reach its peak point ?
E. With what velocity will the mass return to the ground ?
F. What total time is the mass in the air ?

given : $v_o = 29.4$ m/s ; $g = 9.80$ m/s^2

A. the the ball reaches its peak height the velocity will be 0

$v^2 = v_o^2 - 2gy$

$0^2 = (29.4 \text{ m/s})^2 - 2(9.80 \text{ m/s}^2)\, y$; $y = 44.1$ m

B. from part (A) the velocity of the ball is 0.

C. the ball still experiences gravity, therefore the acceleration is still 9.80 m/s^2 ; if the ball did not have an acceleration at the peak point, the ball would remain suspended in midair.

D. $v = v_o - gt$

$0 = 29.4 \text{ m/s} - (9.80 \text{ m/s}^2)\, t$; $t = 3.0$ s

E. the velocity of the ball will be -29.4 m/s when it returns to the ground ; the significance of the (-) sign is to describe the velocity as being downward.

Example 2 A ball is dropped from rest from a building which is 78.4 m tall.
A. How long will it take the ball to strike the ground ?
B. With what velocity will the mass strike the ground ?

given : $v_o = 0$; $g = 9.80$ m/s^2 ; $y = -78.4$ m

A. $y = v_o t - (1/2)gt^2$

$-78.4 \text{ m} = -(1/2)(9.80 \text{ m/s}^2)\, t^2$; $t = 4.00$ s

B. $v = v_o - gt$

$v = 0 - (9.80 \text{ m/s}^2)(4.00 \text{ s}) = -39.2 \text{ m/s}$

Example 3 The ball in Example 2 is now thrown downward with an initial speed of 15.0 m/s.
A. How long will it take the ball to strike the ground ?
B. With what velocity will the mass strike the ground ?

given : $y = -78.4$ m ; $g = 9.80$ m/s^2 ;
$v_o = -15.0$ m/s (negative since the velocity is downward)

Here it may be easier to solve for (b) first then solve for (a). There are other ways to solve this problem correctly !

A. $v^2 = v_{oy}{}^2 - 2gy$

$v^2 = (15.0 \text{ m/s})^2 - 2\,(9.80 \text{ m/s}^2)(-78.4 \text{ m})$

$v = -\,42.0$ m/s (-) since downward

B. $v = v_o - gt$

$-42.0 \text{ m/s} = -15.0 \text{ m/s} - (9.80 \text{ m/s}^2)\,t$; $t = 27.6$ s

Example 4 The ball in Example 3 is now thrown upward with an initial speed of 15.0 m/s.
A. How long will it take the ball to strike the ground ?
B. With what velocity will the mass strike the ground ?
C. How high will the mass ascend relative to the ground ?

given : $y = -78.4$ m ; $g = 9.80$ m/s^2 ; $v_o = 15.0$ m/s

solve for part (B) first : $v^2 = v_o{}^2 - 2gy$

$v^2 = (15.0)^2 - 2(9.80)(-78.4)$; $v = -42.0$ m/s

A. $v = v_o - gt$

$-42.0 \text{ m/s} = 15.0 \text{ m/s} - (9.80 \text{ m/s}^2)\,t$; $t = 5.82$ s

C. at the peak point $v = 0$

$v^2 = v_o{}^2 - 2gy$

$0^2 = (15.0 \text{ m/s})^2 - 2\,(9.80 \text{ m/s}^2)\,y$; $y = 11.5$ m

the total height = 11.5 m + 78.4 m = 89.9 m

Graphs

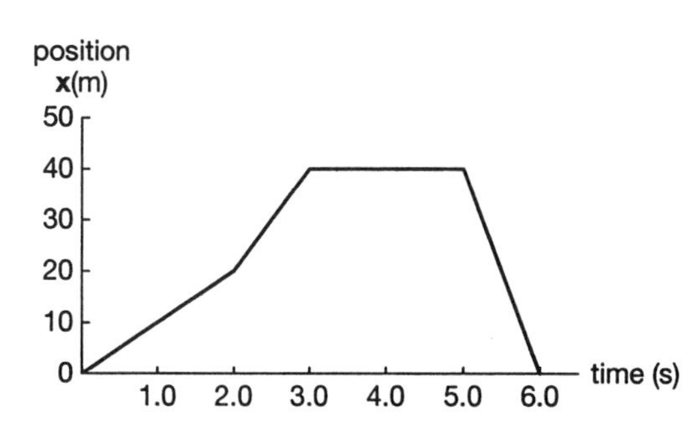

Example 1 The graph above represents the position of a particle vs. time.

A. What is the velocity of the mass at 1.0 s ?
B. What is the velocity of the mass at 2.5 s ?
C. What is the velocity of the mass at 4.0 s ?
D. What is the average velocity of the mass from 0 to 4.0 s ?
E. What is the average velocity for the 6.0 s interval ?

A. the velocity is represented by the slope of the line
$v = \Delta x / \Delta t$
$v = 10\ m / 1\ s = 10\ m/s$
B. velocity is the slope of the line : $v = (40 - 20)\ m / (3.0 - 2.0)\ s = 20\ m/s$
C. the slope of the line is 0.
D. $v = \Delta x / \Delta t$; $v = 40\ m / 4.0\ s = 10\ m/s$
E. $v = 0 / 6.0\ s = 0\ m/s$

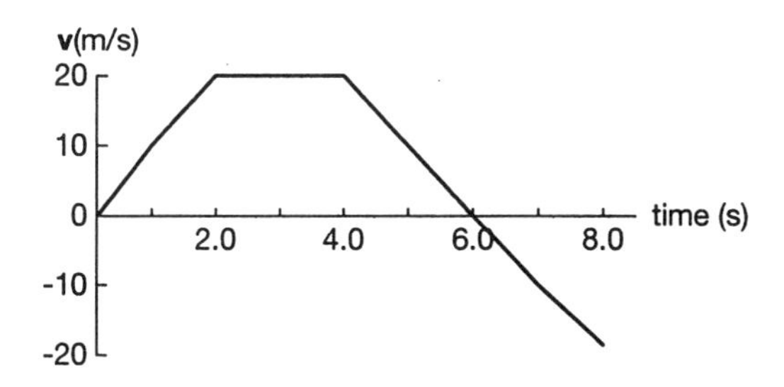

Example 2 The graph above represents the velocity of a mass vs. time.

A. What is the acceleration of the mass at t = 1.0 s ?
B. What is the acceleration of the mass at t = 3.0 s ?
C. What is the average acceleration of the mass between 0 and 5.0 s ?
D. What is the displacement of the mass for the 8.0 s interval ?

A. slope = acceleration ; $a = \Delta v / \Delta t$; $a = (20 - 0)$ m/s / (2.0 s) = 10 m/s^2
B. slope = acceleration ; a = 0
C. $a = \Delta v / \Delta t$; $a = (10 - 0)$ m/s / 5.0 s = 2.0 m/s^2
D. the net area equals the displacement :
Δx_{0-2} = (1/2)(2.0 s) (20 m/s) = 20 m
Δx_{2-4} = (2.0 s)(20 m/s) = 40 m
Δx_{4-6} = (1/2)(2.0 s)(20 m/s) = 20 m
Δx_{6-8} = (1/2)(2.0 s)(-20 m/s) = -20 m
Δx = 20 m + 40 m + 20 m + -20 m = 60 m

Solutions to Selected Problems and Paired Problems

6. (d) equal to the magnitude of the velocity ; the speed can be constant and the velocity change if the particle changes in direction ; example rounding a curve in a car at a speed of 80 km/h

12. given : Δd_1 = 125 km ; Δd_2 = 125 km ; t_1 = 2.0 h ; t_2 = 1.5 h

$s = \Delta d / \Delta t$
(a) s_1 = 125 km / 2.0 h = 63 km/h
s_2 = 125 km / 1.5 h = 83 km/h
(b) s = 250 km / 3.5 h = 71 km/h

16. given : d = (1.50 km)(1000 m/km) = 1.5 x 10^3 m
t = (1.25 min)(60 s / min) = 75 s
$s = \Delta d / \Delta t$
s = (1.5 x 10^3 m) / (75 s) = 20 m/s
the velocity does not have to be constant unless the direction remains the same.

25. given : v_1 = 4.50 m/s ; v_2 = -3.50 m/s ; x_{total} = 100 m

x = vt ; relative velocity $v = v_1 + v_2$; v = 8.00 m/s
100 m = (8.00 m/s) t ; t = 12.5 s

$x_1 = v_1 t$

$x_1 = (4.50\text{ m/s})(12.5\text{ s}) = 56.3\text{ m}$ from runner one's initial position

30. (a) a non-constant acceleration which is positive would be a curve which is curved upward
    (b) a non-constant acceleration which is negative would be a curve which is curved downward
    (c) for (a) x vs. t is concave upward
    for (b) a vs. t is increasing and a concave downward

36. given : $v_o = -2.5\text{ m/s}$; $a = 0.50\text{ m/s}^2$;
    solve for t when v = 0 ; any time greater than t will result in (+) v

    $v = v_o + at$

    $0 = -2.5\text{ m/s} + (0.50\text{ m/s}^2)\,t$; $t = 5.0\text{ s}$; time > 5.0 s

40. (b) since $x = v_o t + (1/2)at^2$; x and t^2 are directly related - general equation for a parabola.

45. given : $v_o = 0$; $a = 5.25\text{ m/s}^2$; $t = 7.00\text{ s}$

    (a) $x = v_o t + (1/2)at^2$

    $x = 0 + (1/2)(5.25\text{ m/s}^2)(7.00\text{ s})^2 = 129\text{ m}$

    (b) $v = v_o + at$

    $v = 0 + (5.25\text{ m/s}^2)(7.00\text{ s}) = 36.8\text{ m/s}$

48. given : $v_o = 25\text{ m/s}$; $v = 0$; $x = 6.0\text{ cm} = 0.060\text{ m}$

    $v^2 = v_o^2 + 2ax$

    $0^2 = (25\text{ m/s})^2 + 2a(0.060\text{ m})$; $a = -5.2 \times 10^3\text{ m/s}^2$

    $v = v_o + at$; $0 = 25\text{ m/s} + (-5.2 \times 10^3\text{ m/s}^2)\,t$; $t = 4.8 \times 10^{-3}\text{ s}$

56. (a) $A = vt$; $x = v_o t + (1/2)at^2$ $(a = 0)$ $x = vt$

    (b) $A = (1/2)at^2$; $x = v_o t + (1/2)at^2$ $(v_o = 0)$ $x = (1/2)at^2$

    (c) $A = v_o t + (1/2)at^2$; $x = v_o t + (1/2)at^2$

60. given : $x_o = 5.0$ m ; $v_o = 10$ m/s ; t = 2.5 s ; x = 65 m

$x = v_{av}t$; 60 m = v_{av}(2.5 m/s) ; $v_{av} = 24$ m/s

$v_{av} = (v + v_o) / 2$; 24 m/s = (v + 10 m/s) / 2 ; v = 38 m/s

$v = v_o + at$; 38 m/s = 10 m/s + a (2.5 s) ; a = 11 m/s^2

70. given : $v_o = 0$; g = 9.80 m/s^2 ; t = 1.80 s

$y = v_ot - (1/2)gt^2$

y = 0 - (1/2)(9.80 m/s^2)(1.80 s)^2 = 15.9 m

76. given : $v_o = -12.4$ m/s ; y = -65.0 m ; g = 9.80 m/s^2

(a) $y = v_ot - (1/2)gt^2$

y = (-12.4 m/s)(2.00 s) - (1/2)(9.80 m/s^2)(2.00 s)^2 = -44.4 m

(b) $v^2 = v_o^2 - 2gy$

v^2 = (-12.4 m/s)^2 - 2(9.80 m/s^2)(-65.0 m)

v = -37.8 m/s

80. given : total time 3.20 s ; $time_{up}$ and $time_{down}$ = 1.60 s ; g = 9.80 m/s^2

(a) $v = v_o - gt$ (at top v = 0)

$0 = v_o$ - (9.80 m/s^2)(1.60 s) ; $v_o = 15.7$ m/s

(b) $v^2 = v_o^2 - 2gy$

0^2 = (15.7 m/s)^2 - 2(9.80 m/s^2) y

y = 12.6 m

88. given : $g_{moon} = g_{Earth} / 6$

(a) $y = v_o t - (1/2)gt^2$; $y = (1/2)gt^2$ (since $v_o = 0$)

y is the same on the moon and the Earth ; set up a proportion

$(1/2)g_m t_m^2 = (1/2)g_e t_e^2$

$(1/6)\, t_m^2 = t_e^2$; $t_m = (2.4)\, t_e$

(b) given : $v_o = 18.0$ m/s

Earth : $v^2 = v_o^2 - 2gy$ $\qquad$ $v = v_o - gt$

$0^2 = (18.0 \text{ m/s})^2 - 2(9.80 \text{ m/s}^2)y$ $\qquad$ $0 = 18.0 \text{ m/s} - (9.80 \text{ m/s}^2)\, t$

$16.5 \text{ m} = y$ $\qquad$ $t_{up} = 1.84$ s ; $t_{total} = 3.68$ s

moon : $0^2 = (18.0 \text{ m/s})^2 - 2(1.63 \text{ m/s}^2)y$ $\qquad$ $0 = 18.0 \text{ m/s} - (1.63 \text{ m/s}^2)\, t$

99.2 m $\qquad$ $11.0 \text{ s} = t_{up}$; $t_{total} = 22.0$ s

Sample Quiz

Multiple Choice. Choose the correct answer.

___ 1. An object moves with a constant speed of 10 m/s. How far will the mass move in 5.0 s ?
A. 0.5 m B. 2.0 m C. 50 m D. 100 m

___ 2. An object moves 5.0 m north and the 3.0 m south. What is the magnitude of distance and displacement of the object ?
A. 2.0 , 8.0 B. 8.0 , 2.0 C. 8.0 , 8.0 D. 8.0 , 15

___ 3. An block, initially at rest, accelerates uniformly. What is the ratio of the velocity of the block after the first second to the velocity after the second second ?
A. 1:1 B. 1:2 C. 1:3 D. 1:4

___ 4. In question 3, what is the ratio of the distance traveled by the object during the first second to the distance traveled during the second second ?
A. 1:1 B. 1:2 C. 1:3 D. 1:4

___ 5. How long will it take a mass to freely fall from a height of 44 m ?
A. 1.0 s B. 2.0 s C. 3.0 s D. 4.0 s

___ 6. An object is thrown upward with a speed of 10 m/s. At the peak point the magnitude of the velocity and acceleration is, respectively,
A. 0 , 0 B. 10 , 0 C. 0 , 9.8 D. 10 , 9.8

___ 7. A ball is thrown vertically upward with a speed v. An identical ball is thrown upward with a speed of 2v. What is the ratio of the height of the second ball to the first ball ?
A. 4:1 B. 2:1 C. 1.4:1 D. 1:1

Problems

8. An airplane travels 300 mi/h south for 2.00 h then 250 mi/h north for 750 miles.
    A. What is the average speed for the trip ?
    B. What is the average velocity for the trip ?

9. A car starts from rest and accelerates for 4.0 m/s^2 for 5.0 s, then maintains that velocity for 10 s and then decelerates at the rate of 2.0 m/s^2 for 4.0 s.
    A. How far did the car move ?
    B. What is the final speed of the car ?

10. A baseball is popped up to the catcher. The time the ball is in the air is 4.5 s.
    A. With what speed did the ball leave the bat ?
    B. How high did the ball ascend ?

CHAPTER 3 Motion in Two Dimensions

Chapter Objectives

Upon completion of the study of motion in two dimensions, students should be able to :

1. work with displacement, velocity, and acceleration vectors so you can
   A. relate acceleration, velocity, displacement, and time for objects in motion with constant velocity or acceleration.
   B. break vectors into components.
   C. add vectors by the following methods :
      (1) triangle method.
      (2) parallelogram method.
      (3) polygon method.
      (4) component method.

2. work with projectile motion so you can
   A. write equations for the x and y components of motion.
   B. solve projectile problems.
   C. sketch graphs of position versus time, velocity versus time, and acceleration versus time for the x and y components of motion.

Chapter Summary

***Curvilinear motion**, or motion in a curved path, has an acceleration at an angle to the instantaneous direction of motion. To analyze such motion in two dimensions, quantities are resolved into rectilinear components : $v_x = v_o \cos\theta$ and $v_y = v_o \sin\theta$. The motion is then considered individually in each linear dimension.

* **Vector Addition** may be done by various graphical methods, including the triangle method and the parallelogram method for two vectors and the polygon method for situations involving more than two vectors. **Vector subtraction** is a special case of vector addition : **A** - **B** = **A** + (-**B**).

*Vector addition is conveniently done by the analytical **component method**.
The recommended procedure is as follows :

(1) Resolve the vectors to be added into their x and y components. Indicate the directions by plus and minus signs.
(2) Add all the x components together and all the y components together to obtain the x and y components of the resultant vector.

(3) Report a vector in component form (unit vector) or in magnitude -- angle form, using the Pythagorean theorem and the arc tangent function.

*For **projectile motion**, the velocity is constant in the horizontal direction, but there is an acceleration (due to gravity) in the vertical direction. The motion is analyzed using components, with time as the common factor.

Important Terms and Relationships

Introduction

Curvilinear motion

3.1 Components of Motion

components of motion

components of velocity : $v_x = v \cos \theta$ and $v_y = v \sin \theta$

components of displacement with zero acceleration : $x = v_x t$ and $y = v_y t$

components of displacement with acceleration : $x = v_{xo} t + (1/2) a_x t^2$

$y = v_{oy} t + (1/2) a_y t^2$

components of velocity : $v_x = v_{xo} + a_x t$ $v_y = v_{yo} + a_y t$

3.2 Vector Addition and Subtraction

vector addition and subtraction
triangle method
parallelogram method
polygon method
component method
magnitude-angle form of a vector
unit vector
component form (of vector)
analytic component method

3.3 Relative Velocity

relative velocity

3.4 Projectile Motion

projectile motion

parabola

range

$x = v_x t$

$v_x = v_o \cos\theta$

$v_{oy} = v_o \sin\theta$

$y = v_{oy}t - (1/2)gt^2$

$v_y^2 = v_{oy}^2 - 2gy$

$v_y = v_{oy} - gt$

$R = v_o^2 \sin 2\theta \ / g$

(only when $y_{initial} = y_{final}$)

Additional Solved Problems

3.1 Components of Motion

Example An airplane has a velocity of 200 m/s , 30° N of E. Find the components of the airplane's velocity

given : $v_o = 200$ m/s ; $\theta = 30°$

$v_x = v \cos\theta$
$v_x = (200 \text{ m/s})(\cos 30°)$
$v_x = 173$ m/s east

$v_y = v \sin\theta$
$v_y = (200 \text{ m/s})(\sin 30°)$
$v_y = 100$ m/s north

3.2 Vector Addition and Subtraction

Example 1 A hiker walks 2.0 miles east then 3.0 miles north, and finally 4.0 miles 30° S of W.

A. What is the position of the hiker relative to his initial position ?

B. In what direction and how far must the hiker walk to return to his original location ?

given : $r_1 = 2.0$ mi east ; $r_2 = 3.0$ mi north ; $r_3 = 4.0$ mi 30° S of W

work with x and y components individually

(a)

$\mathbf{x} = \mathbf{x}_1 + \mathbf{x}_2 + \mathbf{x}_3$
$\mathbf{x} = (2.0 \text{ mi}) + 0 - (4.0 \text{ mi})(\cos 30°)$
$\mathbf{x} = -1.5$ mi or 1.5 mi west

$\mathbf{y} = \mathbf{y}_1 + \mathbf{y}_2 + \mathbf{y}_3$
$\mathbf{y} = 0 + (3.0 \text{ mi}) - (4.0 \text{ mi})(\sin 30°)$
$\mathbf{y} = 1.0$ mi north

$r^2 = x^2 + y^2$
$r^2 = (1.5\ mi)^2 + (1.0\ mi)^2$
$r = 1.8\ mi$

$\tan\theta = 1.0 / 1.5$
$\theta = 34°$ north of west

(b) 1.8 mi 34° south of east

Example 2 Find the resultant of the following vectors.
$v_1 = 3.0$ m/s **x** $v_2 = 8.0$ m/s **y** $v_3 = 12$ m/s 37° above +x

work with x and y components individually

x: $\mathbf{v} = \mathbf{v}_1 + \mathbf{v}_2 + \mathbf{v}_3$
$\mathbf{v} = (3.0\ m/s) + 0 + (12\ m/s)(\cos 37°)$
$\mathbf{v} = 12.6\ m/s$

y: $\mathbf{v} = \mathbf{v}_1 + \mathbf{v}_2 + \mathbf{v}_3$
$\mathbf{v} = 0 + (8.0\ m/s) + (12\ m/s)(\sin 37°)$
$\mathbf{v} = 15.2\ m/s$

one way to express the answer is $\mathbf{v} = (12.6\ m/s)\ \mathbf{x} + (15.2\ m/s)\ \mathbf{y}$

another way is to find the resultant and the angle :

$v^2 = v_x^2 + v_y^2$
$v^2 = (12.6\ m/s)^2 + (15.2\ m/s)^2$; $v = 19.7\ m/s$
$\tan\theta = 15.2 / 12.6$; $\theta = 50°$ above the +x axis

3.3 Relative Velocity

Example 1 A river flows 1.0 m/s south. A boat, whose speed in still water is 5.0 m/s aims directly east across the 150 m wide river.

A. How long does it take the boat to reach the opposite shore ?
B. How far downstream will the boat land ?
C. What is the velocity of the boat as it crosses the river ?

given : $v_r = v_y = 1.0$ m/s south ; $v_x = 5.0$ m/s ; $x = 150$ m

(a) $x = v_x t$
$150\ m = (5.0\ m/s)\ t$; $t = 30\ s$

(b) $y = v_y t$
$y = (1.0\ m/s)(30\ s) = 30\ m$

(c) $v^2 = v_x^2 + v_y^2$
$v^2 = (5.0\ m/s)^2 + (1.0\ m/s)^2$; $v = 5.1\ m/s$

$\tan\theta = (1.0 \text{ m/s}) / (5.0 \text{ m/s})$; $\theta = 11°$ south of east

Example 2 The person in Example 1 wants to travel directly across the river.
A. What angle upstream must the person aim ?
B. How long will it take the person to reach the opposite shore ?
C. With what speed will the boat cross the river ?

(a) $\sin\theta = (1.0 \text{ m/s}) / (5.0 \text{ m/s})$; $\theta = 12°$ north of east

(c) $v^2 = v_x^2 + v_y^2$

$(5.0 \text{ m/s})^2 = v_x^2 + (1.0 \text{ m/s})^2$; $v_x = 4.9 \text{ m/s}$

(b) $x = v_x t$; $150 \text{ m} = (4.9 \text{ m/s})\, t$; $t = 31 \text{ s}$

3.4 Projectile Motion

Example 1 A football is thrown horizontally by a quarterback 1.8 m from the ground with a speed of 15 m/s. Unfortunately the pass is incomplete and the ball strikes the ground.
A. How long will the ball be in the air ?
B. How far from the quarterback will the ball land ?

given : $y = -1.8 \text{ m}$; $v_{oy} = 0$ (since v is horizontal) ; $v_x = 15 \text{ m/s}$; $g = 9.80 \text{ m/s}^2$

first find the time the mass is in the air :

$y = -(1/2)gt^2$

$-1.8 \text{ m} = -(1/2)(9.80 \text{ m/s}^2)\, t^2$

$t = 0.61 \text{ s}$

now find the horizontal distance the ball moves

$x = v_x t$

$x = (15 \text{ m/s})(0.61 \text{ s})$

$x = 9.2 \text{ m}$

Example 2 A soccer ball is kicked. An observer notices the ball travels 20 m horizontally and rises 15 m above the ground.
A. What is the ball's initial horizontal velocity ?
B. What is the ball's initial vertical velocity ?
C. What is the initial velocity of the ball ?

given : $x = 20 \text{ m}$; $y_{max} = 15 \text{ m}$; $g = 9.80 \text{ m/s}^2$

(b) v_y at peak point is 0

$v_y^2 = v_{oy}^2 - 2gy$

$0^2 = v_{oy}^2 - 2\,(9.80\ m/s^2)(15\ m)$

$v_{oy} = 17\ m/s$

(c) now the time the mass is in the air can be found

$v_y = v_{oy} - gt_{up}$

$0 = (17\ m/s) - (9.80\ m/s^2)\,t_{up}$

$t_{up} = 1.7\ s$

$t_{total} = 3.4\ s$

(a) lastly the horizontal velocity may be found

$x = v_x t$

$20\ m = v_x\,(3.4\ s)$; $v_x = 5.9\ m/s$

Example 3 A golfer hits a shot at an angle of 53° above the horizontal with a speed of 30 m/s. The ball lands on an elevated green which is 5.0 m above the point where the golfer hit the ball.
A. How long was the ball in flight ?
B. How far is it from the golfer to the green ?

given : $v_o = 30\ m/s$; $\theta = 53°$; $y = +5.0\ m$; $g = 9.80\ m/s^2$

treat x and y motion independently

$v_x = v_o \cos\theta$

$v_x = (30\ m/s)(\cos 53°) = 18\ m/s$

$v_{oy} = v_o \sin\theta$

$v_{oy} = (30\ m/s)(\sin 53°) = 24\ m/s$

then find v_y when ball strikes the ground

$v_y^2 = v_{oy}^2 - 2gy$

$v_y^2 = (24\ m/s)^2 - 2(9.80\ m/s^2)(5.0\ m)$

$v = -22\ m/s$ (negative since downward)

now find the time ball is in the air

$v_y = v_{oy} - gt$

$-22\ m/s = 24\ m/s - (9.80\ m/s^2)\,t$

$t = 4.7\ s$

now find x

$x = v_x t$

$x = (18\ m/s)(4.7\ s) = 85\ m$

Solutions to selected paired exercises and selected problems

4. The car's velocity and the rain are perpendicular to each other (assuming the rain is falling vertically and the car is traveling horizontally). Both motions must be taken into consideration when finding the apparent velocity. To find the apparent velocity the rusultant must be found.

10. given : $r = 12.5$ cm and $\theta = 210°$

$x = r \cos \theta$ and $y = r \sin \theta$

$x = (12.5 \text{ cm})(\cos 210°)$ $y = (12.5 \text{ cm})(\sin 210°)$

$x = -10.8$ cm $y = -6.25$ cm

16. given : $v_{ox} = 1.5$ m/s ; $a = 0.25 \text{ m/s}^2$ and 37° above the + x axis ; $t = 3.00$ s
treat x and y independently ; first find the x and y components of the acceleration .

$a_x = a \cos 37°$ $a_y = a \sin 37°$

$a_x = (0.25 \text{ m/s}^2)(\cos 37°)$ $a_y = (0.25 \text{ m/s}^2)(\sin 37°)$

$a_x = 0.20 \text{ m/s}^2$ $a_y = 0.15 \text{ m/s}^2$

$x = v_{ox}t + (1/2)a_xt^2$ $y = v_{oy}t + (1/2)a_yt^2$

$x = (1.5)(3.0) + (1/2)(0.20)(3.0)^2$ $y = 0 + (1/2)(0.15)(3.0)^2$

$x = 5.4$ m $y = 0.68$ m

22.
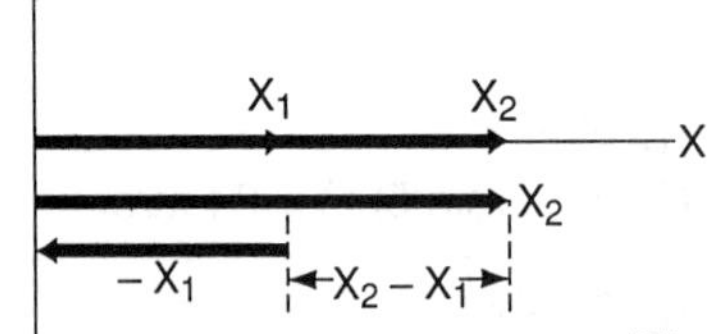

Fig. 3E.22 note : both vectors are in the same direction

26. (a) treat the forces in the **x** direction and forces in the **y** direction independently

F_x : here the forces cancel each other .

F_y : $F_y = F_1 \sin 37° + F_2 \sin 37°$ (add -- same direction)

$F_y = (12.0 \text{ N})(\sin 37°) + (12.0 \text{ N})(\sin 37°) = 14.4$ N

(b) $F_x = F_1 (\cos 27) - F_2 (\cos 37°)$ (subtract - opposite)

$F_x = (12.0 \text{ N})(\cos 27°) - (12.0 \text{ N})(\cos 37°) = 1.1 \text{ N}$

$F_y = F_1 \sin 27° + F_2 \sin 37°$ (add - same direction)

$F_y = (12.0 \text{ N})(\sin 27°) + (12.0 \text{ N})(\cos 37°) = 12.7 \text{ N}$

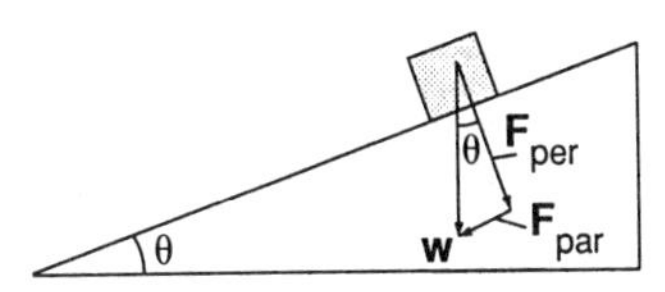

Fig. 3E.26

32. The weight is represented by the hypotenuse in the right triangle. The parallel force is opposite the angle and the perpendicular force is the adjacent side in the right triangle.

$\sin \theta = F_{par} / w$

$\sin 30° = F_{par} / w$

$F_{par} = w / 2$

$\cos \theta = F_{per} / w$

$\cos 30° = F_{per} / w$

$F_{per} = (0.87) w$

35. given : $\mathbf{d}_1 = (3.0 \text{ m}) \mathbf{x} + (3.0 \text{ m})\mathbf{y}$

$\mathbf{d}_2 = (2.5 \text{ m}) \mathbf{x} - (6.0 \text{ m})\mathbf{y}$

$\mathbf{d}_3 = -(2.0 \text{ m})\mathbf{x} + (1.5 \text{ m})\mathbf{y}$

to find $\mathbf{d}_1 + \mathbf{d}_2 + \mathbf{d}_3$ treat x's and y's independent

$x = 3.0 \text{ m} + 2.5 \text{ m} - 2.0 \text{ m} = 3.5 \text{ m}$

$y = 3.0 \text{ m} - 6.0 \text{ m} + 1.5 \text{ m} = -1.5 \text{ m}$

in component notation . $(3.5 \text{ m}) \mathbf{x} - (1.5 \text{ m}) \mathbf{y}$

in magnitude-angle form $d^2 = 3.5^2 + (-1.5)^2$; $d = 3.8$,

$\tan \theta = -1.5 / 3.5$; $\theta = 23°$ below the positive x -- axis

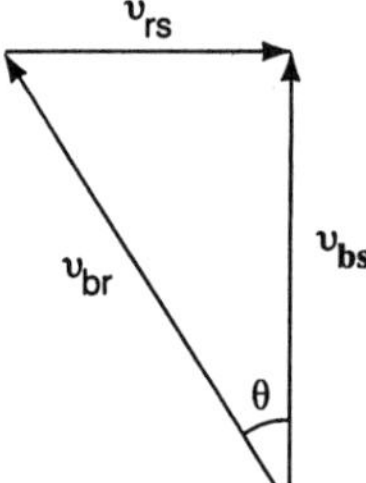

42. Fig. 3E.42

given : v_{br} = 3.50 m/s ; θ =40° ; y = 200 ; x = 25.0 m ; (See Fig. 3E.42)

across river : $y = (v_{br})_y t = (v_{br} \cos 40°)\, t$

$200 \text{ m} = (3.50 \text{ m/s})(\cos 40°)\, t$

$t = 7.46$ s

$x = (v_{br} \sin \theta - v_{rs})\, t$; $x = (v_{br} \sin \theta)\, t - v_{rs}\, t$

$25 \text{ m} = [\,(3.50 \text{ m/s})(\sin 40°)(74.6 \text{ s})] - v_{rs}\,(74.6 \text{ s})$

$v_{rs} = 1.91$ m/s

45. given : v_{bs} = 30.0 km/h = 8.34 m/s ; v_{rs} = 2.50 km/h = 0.695 m/s

(a) $y = v_{br}t$; $y = (8.34 \text{ m/s})(9.00 \text{ s}) = 75.0$ m

$x = v_{rs}t$; $x = (0.695 \text{ m/s})(9.00) = 6.25$ m

(b) $y_{max} = v_{br}t$; $200 \text{ m} = (8.34 \text{ m/s})\, t$; $t = 24.0$ s

$x_{max} = v_{rs}t$; $x_{max} = (0.695 \text{ m/s})(24.0 \text{ s}) = 16.7$ m

48. given : x = 75 m ; x_1 = 25 m ; v_{wg} = 0.30 m/s ; v_{pw} = 0.50 m/s

$v_{pg} = v_{pw} + v_{wg} = 0.80$ m/s

$x_1 = v_{wg}\, t_1$; $25 \text{ m} = (0.30 \text{ m/s})\, t_1$; $t_1 = 83.3$ s

$x - x_1 = v_{pg}t_2$; $50 \text{ m} = (0.80 \text{ m/s})\, t_2$; $t_2 = 62.5$ s

$t = t_1 + t_2 = 146 \text{ s} = 2.43$ min

54. given : $v_x = 1.5$ m/s ; $y = -1.5$ m ; treat x and y motions independently

(a) $y = -(1/2)gt^2$

$-1.5 \text{ m} = -(1/2)(9.8 \text{ m/s}^2)\, t^2$; $t = 0.55$ s

(b) $x = v_x t$; $x = (1.5 \text{ m/s})(0.55 \text{ s}) = 0.83$ m

58. given : $v_x = 23.5$ m/s ; $x = 18.5$ m ; treat x and y motions independently

$x = v_x t$

$18.5 \text{ m} = (23.5 \text{ m/s})\, t$

$t = 0.787$ s

$y = v_{oy}t - (1/2)gt^2$

$y = 0 - (1/2)(9.80 \text{ m/s}^2)(0.787 \text{ s})^2$

$y = 3.03$ m

this is not realistic !

64. $y = v_{oy}t - (1/2)gt^2$ and $x = v_x t$; solve the x equation for t : $t = x / v_x$; substitute into the first equation

$y = v_{yo}(x / v_x) - (1/2)g\,(x / v_x)^2$

$y = (v_y / v_x)\, x - (g / 2v_x^2)(x^2)$; $y = ax - bx^2$

69. given : $v = 0.85$ m/s ; $y = 10$ m ; $t = 2.00$ s

$v_{yo} = v_o \sin\theta$; $v_{yo} = (0.85 \text{ m/s})(\sin 45°) = -0.60$ m/s (-) because downward

$y = v_y t$; $-10 \text{ m} = (-0.60 \text{ m/s})\, t$; $t = 17$ s ; NO

76. given : $y = -4.0$ m (from water) ; $v_o = 8.0$ m/s ; $\theta = 30°$

$v_x = v_o \cos\theta$

$v_x = (8.0 \text{ m/s})(\cos 30)$

$v_x = 6.9$ m/s

$v_{oy} = v_o \sin\theta$

$v_{oy} = (8.0 \text{ m/s})(\sin 30)$

$v_{oy} = 4.0$ m/s

(a) at max height $v_y = 0$: $v_y^2 = v_{oy}^2 - 2gy$

$0^2 = (4.0 \text{ m/s})^2 - 2\,(9.80 \text{ m/s}^2)\, y$; $y = 0.82$ m

height from the water = 0.82 m + 4.0 m = 4.8 m

(b) first find the v_y when the person strikes the water

$v_y^2 = v_{oy}^2 - 2gy$; $v_y^2 = (4.0 \text{ m/s})^2 - 2\,(9.80 \text{ m/s}^2)(-4.0 \text{ m})$; $v_y = -9.7$ m/s

then find the time : $v_y = v_{oy} - gt$; $-9.7 \text{ m/s} = 4.0 \text{ m/s} - (9.80 \text{ m/s}^2)\, t$; $t = 1.4$ s

now find the horizontal distance the person moves : $x = v_x t$

$x = (6.9 \text{ m/s})(1.4 \text{ s}) = 9.7$ m

Sample Quiz

Multiple Choice. Choose the correct answer.

___ 1. The resultant of two vectors is greatest when the angle between them is
A. 0° B. 60° C. 90° D. 180°

___ 2. A boat has a velocity 10 m/s at 53° N of E. What is the north component of the boat's velocity ?
A. 0 m/s B. 6.0 m/s C. 8.0 m/s D. 10 m/s

___ 3. A boat, whose speed in still water is 8.0 m/s, aims directly across a river whose current is 6.0 m/s. What is the speed of the boat as it crosses the river ?
A. 5.3 m/s B. 6.0 m/s C. 8.0 m/s D. 10 m/s

__ 4. A projectile has its greatest range when launched at an angle of __ above the horizontal ?
A. 0° B. 30° C. 45° 60°

question 5 and 6 refer to the following

A ball is launched horizontally from a roof 5.0 m high with a speed of 10 m/s. A second ball is launched horizontally from the same point with a speed of 20 m/s.

__ 5. The time the two balls are in the air are, respectively
A. 1.0 s , 2.0 s B. 2.0 s , 1.0 s C. 1.0 s , 1.0 s D. not listed

__ 6. The horizontal distance the fist ball travels is
A. 5.0 m B. 10 m C. 15 m D. 20 m

__ 7. A soccer ball is kicked with a speed of 20 m/s at an angle of 37° above the horizontal. The speed of the ball at the peak height is
A. B. 12 m/s C. 16 m/s D. 20 m/s

Problems

8. Find the resultant of the following vectors :

$\mathbf{r}_1 = (3.0\text{ m})\,\mathbf{x} + (8.0\text{ m})\,\mathbf{y}$

$\mathbf{r}_2 = (-6.0\text{ m})\,\mathbf{x} - (5.0\text{ m})\,\mathbf{y}$

$\mathbf{r}_3$ = 10 m , 37° above the -x axis

9. A ball thrown horizontally from the top of a building 78.4 m high with a speed of 15 m/s.
   A. How long will the ball be in the air ?
   B. How far from the base of the building will the ball strike the ground ?
   C. With what speed will the ball strike the ground ?

10. A soccer player kicks a ball at 20 m/s at an angle of 53° above the horiontal. The ball lands on the ground down field.
    A. How long is the ball in flight ?
    B. How high will the ball travel ?
    C. How far down field will the ball land ?
    D. With what speed will the ball strike the ground ?
    E. What is the speed of the ball at the peak height ?
    F. What is the acceleration of the ball at the peak height ?

CHAPTER 4 Force and Motion

Chapter Objectives

Upon completion of the unit on force and motion, the students should be able to :

1. define key terms such as dynamics, net force, inertia, and terminal velocity.
2. distinguish between mass and the weight of an object.
3. calculate the weight of an object.
4. state the base units for a newton (N).
5. state Newton's three laws of motion and explain the effect the force has on the motion of a body.
6. describe the effect frictional forces have on bodies.
7. state the two types of friction.
8. list factors which can have an effect upon friction.
9. describe the effects of air resistance.
10. calculate the velocity change which results when a net constant force acts on a body.
11. define and calculate the coefficients of kinetic and static friction.
12. describe a normal force and recognize situations when normal forces are present.
13. when given a force problem, draw a force diagram indicating the forces acting on a body, write equations which relate to Newton's laws of motion, and solve for the unknown(s) in the following situation(s) :
    A. objects at rest.
    B. objects moving with a constant velocity.
    C. objects which are accelerating upward or downward.
    D. objects which are connected by light strings experiencing tensions.
    E. objects on inclined planes.
    F. systems with and without friction.
    G. systems with contact forces.

Chapter Summary

*According to **Newton's first law of motion, the law of inertia**, in the absence of an unbalanced force, a body remains at rest, or if already in motion, the body remains in motion with a constant velocity.

***Inertia** is the natural tendency of an object to maintain a state of rest or at a constant velocity. Mass is a measure of inertia.

*A **force** is something capable of changing an object's state of motion.

* **Newton's second law of motion** states the net force and the acceleration of an object are directly proportional, and the net force and the acceleration are inversely related to each other. It can be written as **F = ma** . The SI unit of force is the newton (N) and 1 N = 1 kg-m/s^2. In the equation, F is the net force and m is the total mass of the system or of any part of it (which may be considered to be a system itself).

*Newton's second law of motion also applies to specific forces, i.e., weight (w = mg).

*__Newton's third law of motion__ states that for every force, there is an equal and opposite reaction force. The opposite force of a force pair act on different objects.

*The magnitude of the **static friction** (f_s) between contacting surfaces is determined by $\mu_s = f_s / N$, where μ_s is the <u>coefficient of static friction</u> and N is the <u>normal force</u>.

*The magnitude of the force of kinetic friction (f_k) is given by $f_k = \mu_k N$, where μ_k is the <u>coefficient of kinetic friction</u> and N is the <u>normal force</u>.

*Air resistance depends on a moving object's shape and size (exposed area) and speed. When the frictional force of a freely falling object equals the weight force, it falls at a constant velocity called the **terminal velocity**.

Important Terms and Relationships

4.1 The Concept of Force ; Net Force

force

4.2 Newton's First Law of Motion

inertia
Newton's first law of motion (law of inertia)

4.3 Newton's Second Law of Motion

Newton's second law of motion F = ma
newton
weight w = mg
free-body diagram

4.4 Applications of Newton's Second Law

translational equilibrium
condition for translational equilibrium

4.5 Newton's Third Law of Motion

Newton's third law of motion

4.6 Friction

force of friction
normal force
static friction
kinetic friction
rolling friction
coefficient of static friction : $f_s \leq \mu_s N$; $f_{s(max)} = \mu_s N$
coefficient of kinetic friction : $f_k = \mu_k N$
air resistance
terminal velocity

Additional Solved Problems

4.3 - 4.4 Newton's Second Law of Motion and Applications

Example 1 A net force of 100 N is applied to an object whose weight is 500 N. What is the acceleration of the mass ?

given : F = 100 N ; w = 500 N

first find the mass : w = mg ; 500 N = m (9.80 m/s^2) ; m = 51.0 kg

now use Newton's 2nd law : F = ma ; 100 N = (51.0 kg) a ; a = 1.96 m/s^2

Example 2 A net force of 10 N is applied to a 2.0-kg mass, initially at rest for a time for 10 s.
A. What is the speed of the mass after the interval of acceleration ?
B. What distance does the mass move during the period of acceleration ?

given : $F_{net} = 10\text{ N}$; $m = 2.0\text{ kg}$

first find the acceleration : $F = ma$; $10\text{ N} = (2.0\text{ kg})\,a$; $a = 5.0\text{ m/s}^2$
constant forces produce constant accelerations :

A. $v = v_o + at$; $v = 0 + (5.0\text{ m/s}^2)(10\text{ s}) = 50\text{ m/s}$

B. $x = v_ot + (1/2)at^2$; $x = 0 + (1/2)(5.0\text{ m/s}^2)(10\text{ s})^2 = 2.5 \times 10^2\text{ m}$

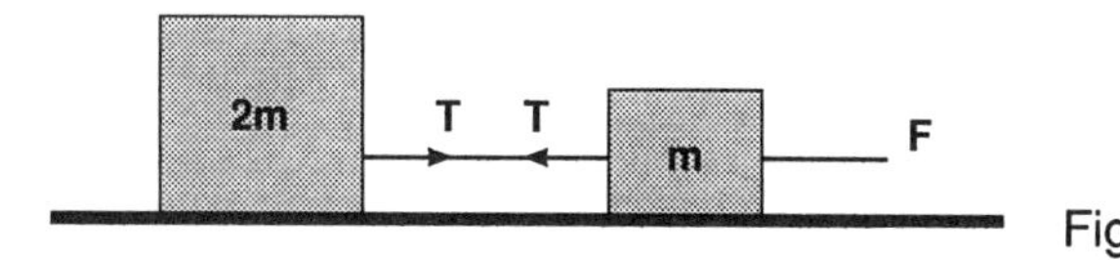

Fig. 4.1

Example 3 In the Fig. 4.1, find the acceleration of the system and the tension in the string between the two masses if $m = 1.0\text{ kg}$ and $F = 12\text{ N}$.

(note only forces in the x should be considered)

for m : $ma = F - T$ and for 2m : $2ma = T$
add the two equations
$ma + 2ma = F$; $3ma = F$; $3\,(1.0\text{ kg})\,a = 12\text{ N}$; $a = 4.0\text{ m/s}^2$
substitue into either expression to find T ; $2(1.0\text{ kg})(4.0\text{ m/s}^2) = T$;
$T = 8.0\text{ N}$

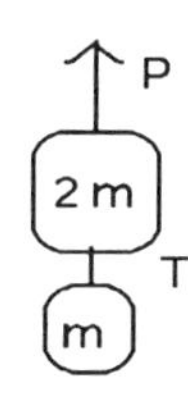

Fig. 4.2

Example 4 In Fig. 4.2, find the acceleration of the system and the tension in the string between the two masses if $m = 1.0\text{ kg}$ and $F = 12\text{ N}$.

top mass : $2ma = F - T - 2mg$ (one upward force and two downward forces)
bottom mass : $ma = T - mg$ (one upward force and one downward force)

add the equations $3ma = F - 3mg$

$3(1.0\text{ kg})(a) = 12\text{ N} - 3\,(1.0\text{ kg})(9.80\text{ m/s}^2)$

$a = -5.8\text{ m/s}^2$

The negative sign means the system is accelerating downward instead of upward. The upward force is less than the total weight of the system.

To find T, substitute into either equation.
(bottom mass) $(1.0\text{ kg})(-5.8\text{ m/s}^2) = T - (1.0\text{ kg})(9.80\text{ m/s}^2)$; $T = 4.0\text{ N}$

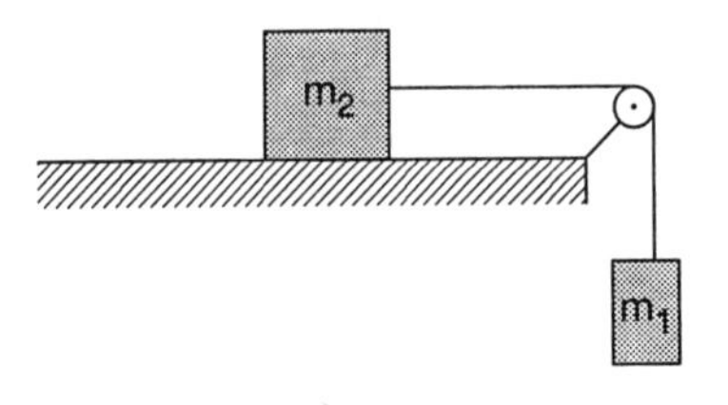

Fig. 4.3

Example 5 In Fig. 4.3, the surface and pulley are ideal. Let $m_1 = 1.0$ kg and $m_2 = 4.0$ kg
A. Find the acceleration of the system.
B. Find the tension in the string connecting the masses.

since the surface is frictionless, the mass m_1 moves downward and m_2 to the right.
for m_1 : $m_1 a = m_1 g - T$ for m_2 : $m_2 a = T$

add the equations : $m_1 a + m_2 a = m_1 g$; $(1.0\text{ kg})a + (4.0\text{ kg})a = (1.0\text{ kg})(9.80\text{ m/s}^2)$;
$a = 2.0\text{ m/s}^2$
(from m_2) $m_2 a = T$; $(4.0\text{ kg})(2.0\text{ m/s}^2) = T$; $T = 8.0\text{ N}$

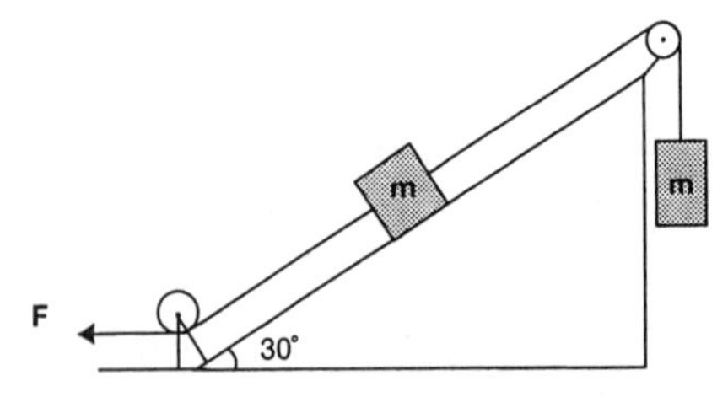

Fig. 4.4

Example 6 In Fig. 4.4, the surface and pulley are ideal. Let m = 1.0 kg and F = 10 N.
A. Find the acceleration of the system.
B. Find the tension connecting the masses.

(a) the mass on the right : $ma = T - mg$
the mass on the inclined plane : $ma = F + mg \sin 30° - T$

add the equations : $2ma = F + mg \sin 30° - mg$

$$2(1.0\ kg)\ a = (10\ N) + (1.0\ kg)(9.80\ m/s^2)(\sin 30°) - (1.0\ kg)(9.80\ m/s^2)$$

$$a = 2.6\ m/s^2$$

(b) to find T subsitute into either equation :

(right) $(1.0\ kg)(2.6\ m/s^2) = T - (1.0\ kg)(9.80\ m/s^2)$; $T = 12.4\ N$

4.6 Friction

Example 1 A 5.0 kg mass slides on a horizontal surface. The coefficient of kinetic friction between the surfaces is 0.25. Find the force needed to keep the object in motion with a constant speed.

given : $\mu_k = 0.25$; $m = 5.0\ kg$

the force needed to keep the mass moving with a constant speed must equal the force of kinetic friction ; since the mass is on a horizontal surface the normal force equal the object's weight.

$f_k = \mu_k N$; $f_k = \mu_k mg$; $f_k = (0.25)(5.0\ kg)(9.80\ m/s^2) = 12\ N$

Example 2 Repeat Example 3 in the previous section if the surfaces have a coefficient of kinetic friction of 0.25.

(a) the only change is that friction impedes the motion.

for mass on the left : $2ma = T - f_k$; $f_k = \mu_k 2mg$; $2ma = T - \mu_k(2\ mg)$
for mass on the right : $ma = F - T - f$; $f_k = \mu_k mg$; $ma = F - T - \mu_k mg$
adding equations : $3ma = F - \mu_k(3mg)$

$$3(1.0\ kg)\ a = 12\ N - (0.25)(3)(1.0\ kg)(9.80\ m/s^2)$$

$$a = 1.6\ m/s^2$$

(b) substituting : $2(1.0\ kg)(1.6\ m/s^2) = T - (0.25)(2)(1.0\ kg)(9.80\ m/s^2)$; $T = 8.1\ N$

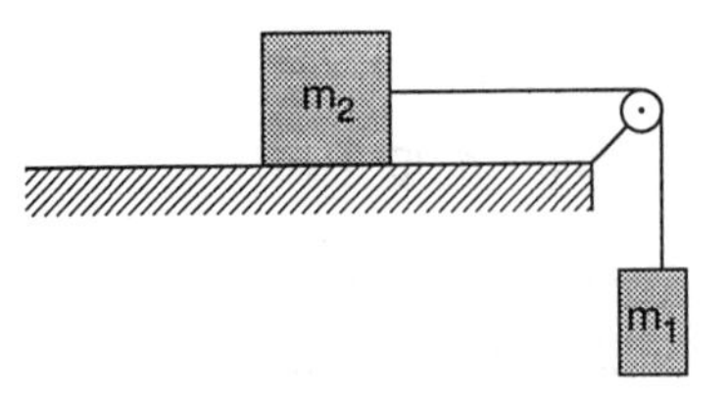

Example 3

Fig. 4.5

In the system shown in Fig. 4.5 $m_1 = 3.0$ kg and $m_2 = 4.0$ kg. Find the minimum value for μ_s needed to keep the system at rest.

since the system is at rest the acceleration is 0.

mass on right : $0 = m_1g - T$

mass on surface : $0 = T - f_s$; $f_s = m_1g$; $f_s = (3.0 \text{ kg})(9.80 \text{ m/s}^2) = 29$ N

$f_s = \mu_s mg$; $29 \text{ N} = \mu_s (4.0 \text{ kg})(9.80 \text{ m/s}^2)$; $\mu_s = 0.74$

Example 4 A mass m moves down a rough 37° inclined plane ($\mu_k = 0.20$).

A. Find the acceleration of the mass as it moves down the plane.
B. If the plane is 10 m long and the mass starts from rest, what will be its speed at the bottom of the plane ?

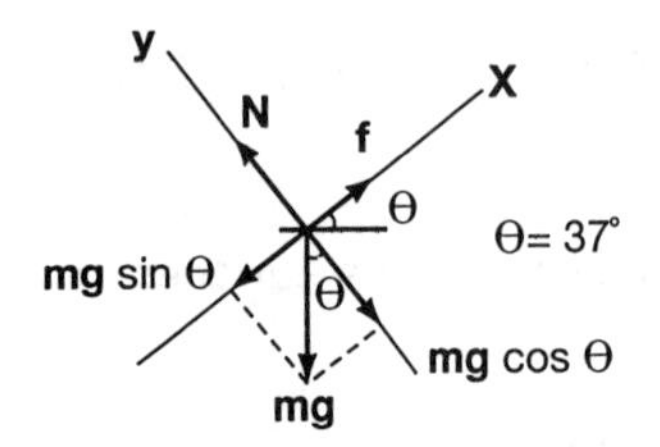

Fig. 4.6

(a) $ma = mg \sin\theta - f_k$; $f_k = \mu_k N$; $N = mg\cos\theta$

$ma = mg\sin\theta - \mu_k N$; $ma = mg\sin\theta - \mu_k mg\cos\theta$

$a = (9.80 \text{ m/s}^2)(\sin 37°) - (0.20)(9.80 \text{ m/s}^2)(\cos 37°) = 4.3 \text{ m/s}^2$

(b) $v^2 = v_o^2 + 2ax$; $v^2 = 0^2 + 2(4.3 \text{ m/s}^2)(10 \text{ m})$; $v = 9.3$ m/s

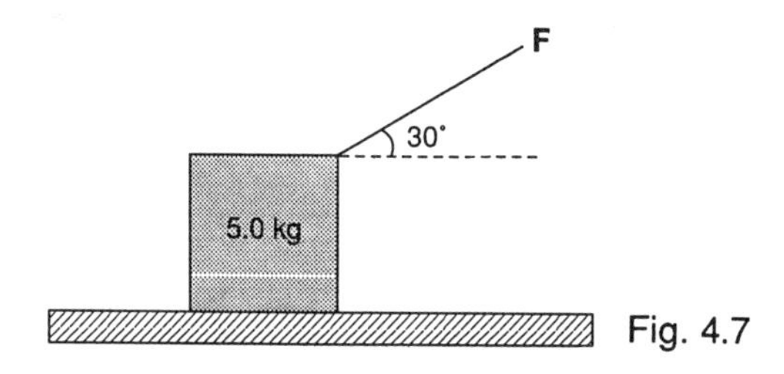

Fig. 4.7

Example 5 For the situation in Fig. 4.7 find the force F needed to pull the mass with a constant speed. The coefficient of kinetic friction between the movable surfaces is 0.20.

there are four forces acting on the mass : F , f_k , N , and mg
since the object is moving with a constant speed, the acceleration is zero
first write the equation for forces in the x : $0 = F(\cos 30°) - f_k$
second write the equation for forces in the y : $0 = N + F(\sin 30°) - mg$
there are three unknows in the two equations, therefore we need another equation.

$f_k = \mu_k N$; take the original equations, solve for f_k and N and substitute.
$F(\cos 30°) = \mu_k [mg - F(\sin 30°)]$

$F(0.87) = (0.20)[(5.0\text{ kg})(9.80\text{ m/s}^2) - F(0.5)]$
$(0.87)F = 9.8 - 0.1F$
$0.97F = 9.8$; $F = 10\text{ N}$

Solutions to paired problems and selected problems.

2. (d) this question is a little tricky. Both (a) and (b) are correct responses according to Newton's first law of motion.

10. since the object moves with a constant velocity, the net force is zero

find the resultant force for $F_1 + F_2$
x : $F_{1x} + F_{2x} = (5.0\text{ N})(\cos 30°) - (3.0\text{ N})(\cos 37°) = 1.9\text{ N}$
y : $F_{1y} + F_{2y} = (5.0\text{ N})(\sin 30°) - (3.0\text{ N})(\sin 37°) = 0.70\text{ N}$
the third force should have components which can be added to $\mathbf{F}_1 + \mathbf{F}_2$ to equal zero
$\mathbf{F}_3 = (-1.9\text{ N})\,\mathbf{x} - (0.70\text{ N})\,\mathbf{y}$

18. given : $F_{net} = 100$ N ; $a = 1.50$ m/s^2

$F = ma$

$100 \text{ N} = (m)(0.75 \text{ m/s}^2)$

$m = 1.3 \times 10^2$ kg

$w = mg$

$w = (1.3 \times 10^2 \text{ kg})(9.80 \text{ m/s}^2)$

$w = 1.3 \times 10^3$ N

26. given : $F = 300$ N ; $m = 7.50$ kg ; $f_k = 65.0$ N

$ma = F - f_k$

$(7.50 \text{ kg})\, a = (300 \text{ N}) - (65.0 \text{ N})$

$a = 31.3$ m/s^2

28. given : $v_o = 50$ km / h $= 14$ m/s ; $v = 15$ km/h $= 4.2$ m/s ; $t = 3.0$ s ; $m = 65$ kg

first find the acceleration : $v = v_o + at$

$4.2 \text{ m/s} = (14 \text{ m/s}) + a\,(3.0)$

$a = -3.3$ m/s^2

now the net force can be found : $F = ma$

$F = (65 \text{ kg})(-3.3 \text{ m/s}^2) = -2.1 \times 10^2$ N

32. given : $w = 2.75 \times 10^6$ N ; $F_{thrust} = 6.35 \times 10^6$ N ; $v_o = 0$;
$v = 285$ km/h $= 79.2$ m/s

first find the mass of the plane : $w = mg$; $2.75 \times 10^6 \text{ N} = m\,(9.80 \text{ m/s}^2)$

$m = 2.81 \times 10^5$ kg

now find the acceleration : $F = ma$; $6.35 \times 10^6 \text{ N} = (2.81 \times 10^5 \text{ kg})\, a$

$a = 22.6$ m/s^2

now work with constant acceleration : $v^2 = v_o{}^2 + 2ax$

$(79.2 \text{ m/s})^2 = 0^2 + 2(22.6 \text{ m/s}^2)\, x$

$x = 139$ m

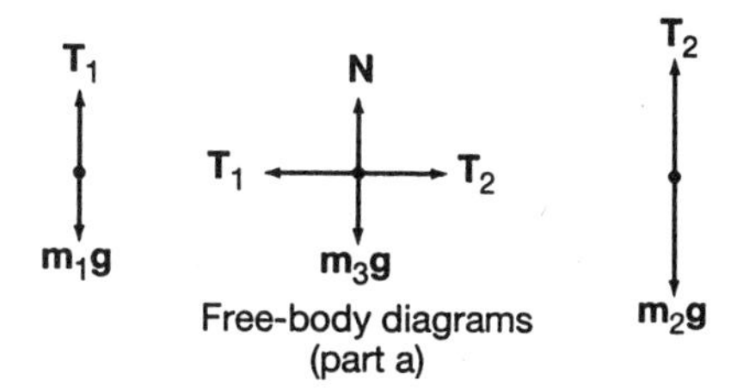

Free-body diagrams
(part a)

Fig. 4E.45

45. Sketch the free-body diagram for each body before writing the equations. The mass m_2 moves downward since its mass is greater than m_2. (Vertical forces on m_3 cancel.)

$m_1a = T_1 - m_1g \qquad m_3a = T_1 - T_2 \qquad m_2a = m_2g - T_2$

Now add the equations :

$m_1a + m_2a + m_3a = m_2g - m_1g$

(a) (0.25 kg)a+ (0.25 kg)a + (0.50 kg)a = (0.50 kg)(9.80 m/s^2) - (0.25 kg)(9.80 m/s^2)
a = 2.5 m/s^2, right.

(b) the only difference are the masses ; the system moves in opposite direction
(0.35 kg)a+ (0.15 kg)a + (0.50 kg)a = (0.50 kg)(9.80 m/s^2) - (0.35 kg)(9.80 m/s^2)
a = 2.0 m/s^2, left.

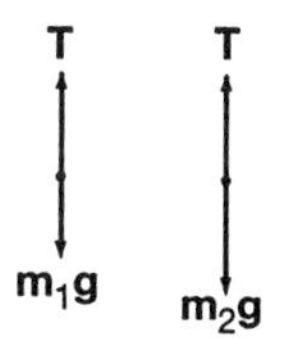

Free-body diagrams

Fig. 4E.48

48. given : m_1 = 215 g = 0.215 kg ; m_2 = 255 g = 0.255 kg ; $y_{\text{to floor}}$ = 1.10 m
see free-body diagrams in Fig. 4E.48

(a) $m_1a = T - m_1g \qquad m_2a = m_2g - T$

add the two equations

$m_1a + m_2a = m_2g - m_1g$

(0.215 kg) a + (0.255) a = (0.255 kg)(9.80 m/s^2) - (0.215 kg)(9.80 m/s^2)
a = 0.83 m/s^2 ; y = (1/2)at^2 ; 1.10 m = (1/2)(0.83 m/s^2) t^2 ; t = 1.6 s

(b) find the speed of the system when m_2 strikes the floor :

$v^2 = v_0^2 + 2ay$; $v^2 = 0^2 + 2$ (0.83 m/s^2)(1.10 m) ; v = 1.35 m/s

the height of m_1 at that point is 1.10 m ; the mass continues to move upward since its velocity is $\neq 0$.

$v^2 = v_o^2 - 2gy$; $0^2 = (1.35 \text{ m/s})^2 - 2\,(9.80 \text{ m/s}^2)\, y_2$; $y_2 = 0.093$ m

the total height then is 1.19 m

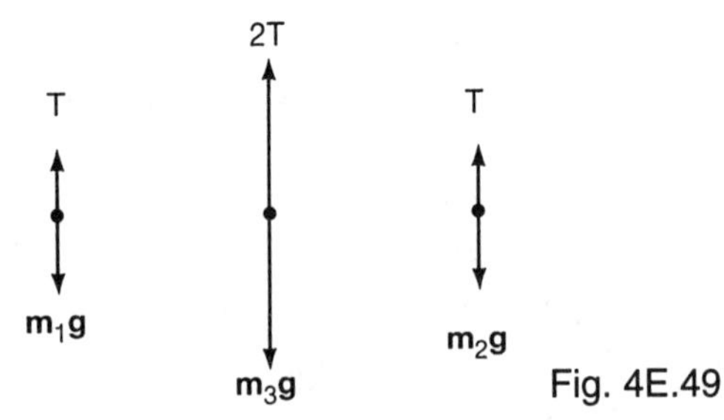

Fig. 4E.49

49. sketch free-body diagrams on the masses before attempting to write the equations (see Fig. 4E.49)

(a) $m_1a = m_1g - T$ $\quad$ $m_3a = 2T - m_3g$ $\quad$ $m_2a = m_2g - T$

add the equations together

$$m_1a + m_3a + m_2a = m_1g + m_2g - m_3g$$

$$(3.0 \text{ kg})a + (4.0 \text{ kg})a + (3.0 \text{ kg})a = (3.0 \text{ kg})(9.80 \text{ m/s}^2) + (3.0 \text{ kg})(9.80) - (4.0)(9.8)$$

$$a = 2.0 \text{ m/s}^2$$

(b) substitute into either equation to find T

$$(3.0 \text{ kg})(2.0 \text{ m/s}^2) = (3.0 \text{ kg})(9.80 \text{ m/s}^2) - T \ ; \ T = 23 \text{ N}$$

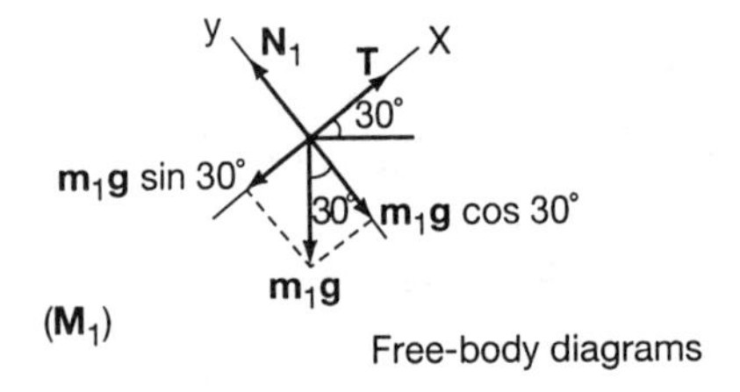

Free-body diagrams

Fig. 4E.52

52. m_2 moves down the plane since its mass is much larger and the degree of inclined plane is larger.

$$m_1a = T - m_1g \sin\theta \qquad m_2a = m_2g \sin\theta - T$$

add the equations

$$m_1a + m_2a = m_2g \sin\theta - m_1g \sin\theta$$

$$(15 \text{ kg})a + (25 \text{ kg})a = (25 \text{ kg})(9.80 \text{ m/s}^2)(\sin 37°) - (15 \text{ kg})(9.80)(\sin 30°)$$

$$a = 1.8 \text{ m/s}^2$$

58. since the masses are not given , we do not know if the system will move or accelerate.

for m_1 : the weight acts downward and the tension in the string acts upward

for m_2 : the weight acts downward and the tension in the string acts upward

for m_3 : the table pushes up on the mass to counteract the weight of m_3 : the tensions in the two string act in opposite directions

66. in both situations minimizing air resistance is done by reducing the surface area for oncoming air

68. (a) a force to overcome static friction must be present to start the mass moving.

$f_s = 275\ N$

$f_s = \mu_s\ N$

(since the surface is horizontal, the normal force equals the weight of the desk)

$275\ N = \mu_s\ (35.0\ kg)(9.80\ m/s^2)$

$\mu_s = 0.802$

(b) once the object is in motion the force needed to keep the desk in motion equals the force of kinetic friction.

$f_k = \mu_k mg$

$195\ N = \mu_k\ (35.0\ kg)(9.80\ m/s^2)$; $\mu_k = 0.569$

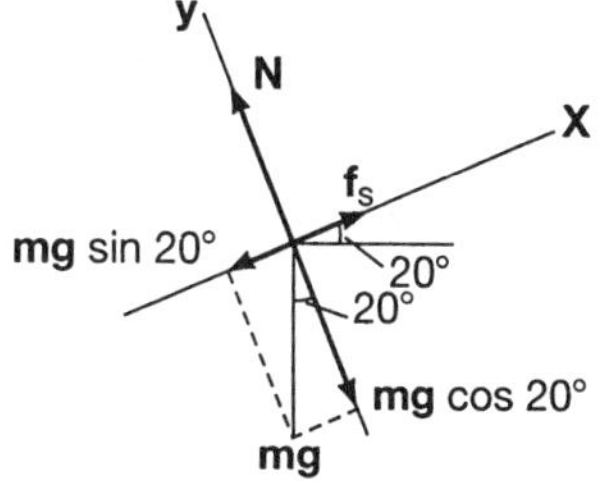

Fig. 4E.72

72. first, find the maximum value for static friction (see free-body diagram in Fig4E.72)

$f_s = \mu_s N$

the normal force does not equal the weight but the perpendicular component of the weight to the surface. $N = mg \cos \theta$ (see Fig. 4E.72)

$f_s = \mu_s mg \cos \theta$

$f_s = (0.65)(m)(9.80\ m/s^2)(\cos 20°) = (5.96)\ m$

the parallel component of the weight down the inclined plane is

$F_{par} = mg \sin\theta$; $F_{par} = (m)(9.80\ m/s^2)(\sin 20°) = (3.35)\ m$

since the value for the static friction exceeds the force pulling the mass down the plane, the object will remain at rest.

80. first find the force of static friction : $f_s = \mu_s N$

since the car is on level ground, the normal force equals the car's weight.

$f_k = \mu_k mg$

(a) $f_k = (0.85)(1500\ kg)(9.80\ m/s^2) = 1.25 \times 10^4\ N$

$ma = f$; $(1500)(a) = 1.25 \times 10^4\ N$; $a = 8.3\ m/s^2$

$v^2 = v_o^2 - 2ax$ (90 km /h = 25 m/s)

$0^2 = (25\ m/s)^2 - 2(8.3)x$ $x = 38\ m$

(b) $f_s = (0.60)(1500\ kg)(9.80\ m/s^2) = 8.8 \times 10^3\ N$

$(1500\ kg)\ a = (8.8 \times 10^3\ N)$; $a = 5.9\ m/s^2$

$0^2 = (25\ m/s)^2 - 2\ (5.9\ m/s^2)\ x$; $x = 53\ m$

86 (a) $f_k = m_2 g - m_1 g = (0.250)(9.80) - (0.150)(9.80) = 0.980\ N$

$\mu_k = f_k / N$; $m_3 = (0.980) / (0.560)(9.80) = 0.179\ kg$

(b) $m_2 a = m_2 g - T_1$; $m_3 a = T_1 - T_2 - \mu_k m_3 g$; $m_1 a = T_2 - m_1 g$

$m_1 a + m_2 a + m_3 a = m_2 g - m_1 g - \mu_k m_3 g$

$(0.500)a = (0.250)(9.80) - (0.150)(9.80) - (0.560)(0.100)(9.80)$

$a = 0.860\ m/s^2$

Sample Quiz

Multiple Choice. Choose the correct answer.

___ 1. A force F accelerates a mass m with an acceleration a. If the same force is applied to mass (m/2) the acceleration will be
A. a / 4 B. a / 2 C. 2 a D. 4 a

___ 2. A body is acted on by two force, one 4.0 N in the **+x** direction and a 3.0 N force in the **-y** direction. What single force could be added to produce equilibium ?
A. 5.0 N , 37° below +x axis C. 7.0 N , 37° below +x axis
B. 5.0 N , 37° above -x axis D. 7.0 N , 37° above -x axis

___ 3. An object has a mass of 5 g. What is the approximate weight of the mass ?
A. 50 N B. 5 N C. 0.005 N D. 0.05 N

___ 4. Two vectors of magnitudes 6.0 N and 8.0 N are added. What is the maximum possible resultant of the vectors ?
A. 2.0 N B. 10 N C. 14 N D. 48 N

___ 5. A person having a mass of 50 kg rides in an elevator. His apparent weight on a set of scales reads 600 N. What is his approximate acceleration ?
A. 12 m/s^2 up B. 12 m/s^2 down C. 2.0 m/s^2 up D. 2.0 m/s^2 down

___ 6. Which of Newton's laws of motion best explains why motorists should buckle-up ?
A. first law B. second law C. third law D. gravitation

___ 7. A mass slides down an inclined plane, angle θ, with a constant speed. The value for the coefficient of kinetic friction μ_k is
A. cos θ B. sin θ C. sec θ D. tan θ

Problems

8. A boy pulls a 15 - kg sled with a horizontal force of 20 N on a icy surface (frictionless). Find the acceleration of the sled.

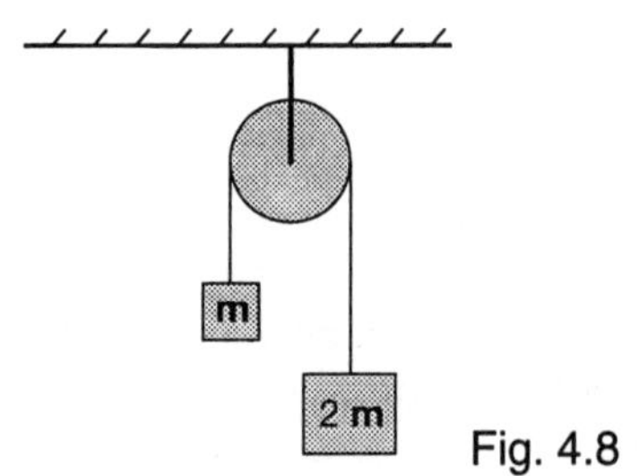

Fig. 4.8

9. In Fig. 4.8, find the acceleration of the system and the tension in the connecting strings. m = 2.0 kg (neglect friction and mass of the pulley)

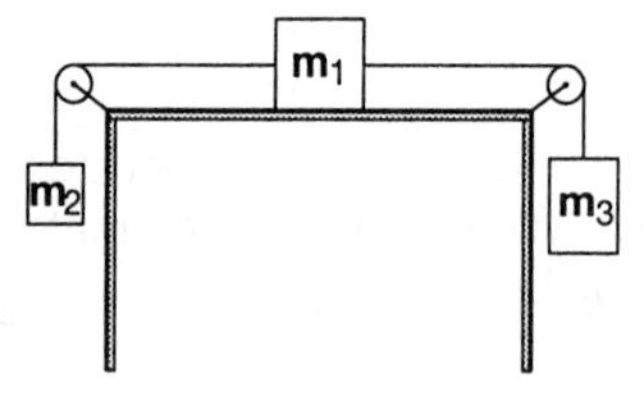

Fig. 4.9

10. In Fig. 4.9 , m_1 = 5.0 kg , m_2 = 2.0 kg , and m_3 = 4.0 kg. The coefficient of kinetic friction betweenm$_1$ and the table is 0.20. Find the acceleration of the system. (Neglect friction and rotational effects of the pulley.)

CHAPTER 5 Work and Energy

Chapter Objectives

Upon completion of the unit on work, energy, and power, students should be able to :

1. define work, energy, and power.
2. express a joule and a watt in their base units.
3. calculate the work done on a body when the force is constant and the body undergoes a specified displacement.
4. calculate the work by finding the area under a graph of force as a function of position.
5. write the expression and solve problems for kinetic energy.
6. relate the net work done with the change in kinetic energy.
7. understand potential energy as energy of position ; write the expression for potential energy for a body in a uniform graviational field and the expression for the potential energy for a spring.
8. distinguish between conservative and non-conservative forces.
9. identify situations where mechanical energy is or is not conserved and solve problems on the law of conservation of energy and the loss of energy.
10. write the equation for power and solve problems on power.
11. identify and write expressions for the power when an object moves with a constant speed due to a constant force.
12. write and apply the expression for efficiency in terms of work and power.

Chapter Summary

*The **work** done by a constant force in moving an object is equal to the product of the magnitude of the displacement and the component of the force parallel to the displacement. The SI unit of work is the joule (J).

*For a variable spring force as given by Hooke's law, $F = kx$, the work done in stretching (or compressing) a spring is equal to $W = (1/2)\ kx^2$.

***Kinetic energy** (K) is the energy of motion. The **work-energy theorem** states that the net work on an object is equal to its change in kinetic energy. In general, work is a measure of energy transfer.

*$\textbf{Potential energy}$ (U) is the energy of position. For example, as a result of work being done in compressing a spring, there is a change in position and a change in potential energy (ΔU). With $x_o = 0$, then $U = (1/2)kx^2$. Similarly, for gravitational potential energy, with $h_o = 0$, $U = mgh$.

*Systems may be **conservative** or **nonconservative**. A conservative system is one if the work done by the force is independent of the object's path. The force of gravity is an example of a conservative force. Friction is an example of a non-conservative force, which is path dependent.

*The total **mechanical energy** of a system is the sum of its potential and kinetic energies ($E = K + U$). The total mechanical energy is conserved in a conservative system.

***Power** is the rate of doing work. The SI unit of power is the watt (W), and the British unit is ft-lb/s. A common British unit of power is the horsepower (hp). (1 hp = 550 ft -lb/s = 746 W).

***Efficiency** is a comparison of the useful work output to the energy input given as a fraction (or percentage).

Important Terms and Relationships

5.1 Work Done by a Constant Force

work
joule
$W = Fd \cos \theta$

5.2 Work done by a variable force

area for F vs. d represents work
Hooke's law : $F = -kx$
work done on a spring : $W = (1/2)\, kx^2$
spring constant

5.3 The Work-Energy Theorem : Kinetic Energy

work- energy theorem : $W_{net} = K - K_o = \Delta K$
kinetic energy : $K = (1/2)mv^2$

5.4 Potential Energy

potential energy
gravitational : $U = mgh$
spring : $U = (1/2)kx^2$

5.5 The Conservation of Energy

Conservation of Mechanical Energy : $K_1 + U_1 = K_2 + U_2$
Conservation of total energy
conservative system
total mechanical energy
Conservation of Energy with a non-conservative force $K_1 + U_1 = K_2 + U_2 + Q$

5.6 Power

power : $P = W / t$
watt (W)
horsepower (hp)
efficiency : $\varepsilon = (W_{out} / E_{in}) \times 100\%$ or $(P_{out} / P_{in}) \times 100\%$

Additional Solved Problems

5.1 Work

Example 1 A 5000 N elevator is pulled upward with a force of 6000 N for a distance of 100 m.
A. Find the work done by the upward force.
B. Find the work done by the gravitational force.
C. Find the work done by the net force.

given : $w = 5000$ N ; $F_{up} = 6000$ N ; $d = 100$ m

(a) $W_{up} = F d \cos\theta$; $W_{up} = (6000 \text{ N})(100 \text{ m})(\cos 0°) = 6.00 \times 10^6$ J

(b) $W_{grav} = mg\, d \cos\theta$; $W_{grav} = (5000 \text{ N})(100 \text{ m})(\cos 180°) = -5.00 \times 10^6$ J

(c) $W_{net} = W_{up} + W_{grav}$; $W_{net} = (6.00 \times 10^6 \text{ J}) + (-5.00 \times 10^6 \text{ J})$

$W_{net} = 1.00 \times 10^6$ J

Example 2 A 2.0-kg mass slides down a 30° inclined plane 10 m long with constant speed.
A. Find the work done by the gravitational force.
B. Find the work done by the frictional force.
C. Find the net work done.

given : $m = 2.0$ kg ; $\theta = 30°$; $d = 10$ m

(a) $W_{grav} = mg\,d\cos\theta$;

$W_{grav} = (2.0\text{ kg})(9.80\text{ m/s}^2)(10\text{ m})(\cos 60°) = 98\text{ J}$

(b) $W_{friction} = f_k\,d\cos\theta$;

since object moves with constant speed $f_k = mg\sin 30°$

$f_k = (2.0\text{ kg})(9.80\text{ m/s}^2)(\sin 30°) = 9.8\text{ N}$

$W = (9.8\text{ N})(10\text{ m})(\cos 180) = -98\text{ J}$

(c) $W_{net} = W_{mg} + W_{friction} = 98\text{ J} + (-98\text{ J}) = 0$

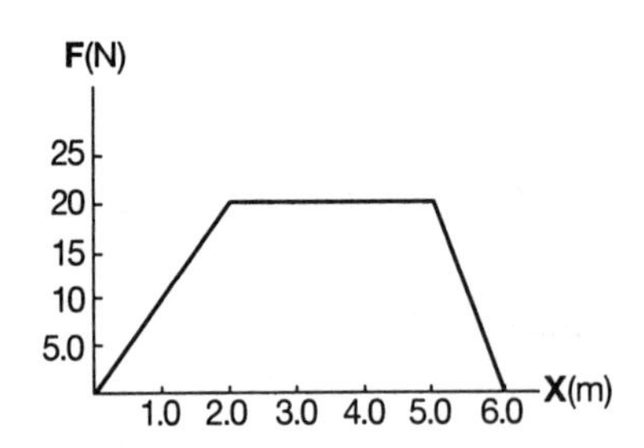

Example 3 The graph above show the force on a mass vs. the object's parallel displacement. Find the work done when the mass moves from 0 to 6.0 m.

to find the work done, find the area for the graph ; separate the problem in three parts : from 0 - 2.0 m , 2.0 m - 5.0 m, and 5.0 m to 6.0 m.

$W_{0-2} = (1/2)(20\text{ N})(2.0\text{ m}) = 20\text{ J}$

$W_{2-5} = (20\text{ N})(3.0\text{ m}) = 60\text{ J}$

$W_{5-6} = (1/2)(20\text{ N})(1.0\text{ m}) = 10\text{ J}$; $W_{total} = 90\text{ J}$

5.3 Work-Energy Theorem

Example 1 A 5.0 - kg mass at rest is acted upon a force which does 40 J of work. What is the final speed of the mass ?

given : $m = 5.0$ kg ; $W = 40$ J
the net work done equals the change in kinetic energy :

$W = \Delta K$; $40 \text{ J} = \Delta K$; $\Delta K = K - K_o$; $40 \text{ J} = K - 0$; $K = 40$ J

$K = (1/2)\, mv^2$; $40 \text{ J} = (1/2)(5.0 \text{ kg})\, v^2$; $v = 4.0$ m/s

Example 2 A 0.25-kg mass is moving at a speed of 20 m/s when the net work done on the mass is -20 J. What is the new speed of the mass ?

given : $m = 0.25$ kg ; $v_o = 20$ m/s ; $W_{net} = -20$ J

first find the initial kinetic energy $K_o = (1/2)mv_o{}^2 = (1/2)(0.25 \text{ kg})(20 \text{ m/s})^2$

$K_o = 50$ J

$W = \Delta K$; $W = K - K_o$; $-20 \text{ J} = K - 50 \text{ J}$; $K = 30$ J

$K = (1/2)\, mv^2$; $30 \text{ J} = (1/2)(0.25 \text{ kg})(v^2)$; $v = 15$ m/s

5.4 Potential Energy

Example 1 A 60-kg person stands on a diving board overlooking a pool of water. The board is 10 m above the water. What is the potential energy of the the diver with respect to the water ?

given : $m = 60$ kg ; $h = 10$ m
gravitational potential energy : $U = mgh$

$U = (60 \text{ kg})(9.80 \text{ m/s}^2)(10 \text{ m}) = 5.9 \times 10^3$ J

Example 2 A spring has a force constant of 150 N/m.

A. What is the potential energy of the spring when it is compressed 20 cm ?

B. What is the potential energy of the spring when it is compressed 40 cm ?

given : $k = 150$ N/m ; $x_1 = 0.20$ m ; $x_2 = 0.40$ m

elastic potential energy : $U = (1/2)kx^2$

(a) $U = (1/2)(150 \text{ N/m}) (0.20 \text{ m})^2 = 3.0 \text{ J}$

(b) $U = (1/2)(150 \text{ N/m})(0.40 \text{ m})^2 = 12 \text{ J}$

5.5 The Conservation of Energy

Example 1 A pendulum 1.0 m long is released when the string is horizontal.
A. What is the speed of the mass when it reaches the bottom of its path ?
B. What is the speed of the mass when the string makes an angle of 30° with the vertical ?

(a) when released from the horizontal (taking the bottom as the zero height)

$$\Delta h = 1.0 \text{ m} \ ; \quad U_1 + K_1 = U_2 + K_2$$

$$mgh_1 + 0 = 0 + (1/2)mv_2^2$$

$$m\,(9.80 \text{ m/s}^2)(1.0 \text{ m}) = (1/2)\,m\,(v_2^2)$$

$$v_2 = 4.4 \text{ m/s}$$

(b) $\Delta h = L - L\cos\theta$; $\Delta h = 1.0 \text{ m} - (1.0 \text{ m})(\cos 30°) = 0.13 \text{ m}$

$$U_1 + K_1 = U_3 + K_3$$

$$m\,(9.80 \text{ m/s}^2)(1.0 \text{ m}) + 0 = m\,(9.80 \text{ m/s}^2)(0.13 \text{ m}) + (1/2)\,m\,v_2^2$$

$$v_2 = 4.1 \text{ m/s}$$

Example 2 A 0.50-kg ball is thrown with a speed of 10 m/s from the top of a building 20 m above the ground.
A. What is the potential and kinetic energy of the mass as it leaves the person's hand ?
B. What is the potential and kinetic energy of the mass as it strikes the ground ?
C. What is the speed of the mass as it strikes the ground ?

given : $m = 0.50$ kg ; $v_o = 10$ m/s ; $h = 20$ m

(a) the potential energy $U = mgh$; $U = (0.50 \text{ kg})(9.80 \text{ m/s}^2)(20 \text{ m}) = 98 \text{ J}$
the kinetic energy $K = (1/2)mv^2$; $K = (1/2)(0.50 \text{ kg})(10 \text{ m/s})^2 = 25 \text{ J}$

(the total energy equals K + U = 123 J)

(b) $U_1 + K_1 = U_2 + K_2$

$98\text{ J} + 25\text{ J} = 0 + K_2$; $K_2 = 123\text{ J}$

(c) $K_2 = (1/2)mv_2^2$; $123\text{ J} = (1/2)(0.50\text{ kg})(v_2^2)$; $v_2 = 22\text{ m/s}$

Example 3 A 0.20-kg mass moves with a speed of 3.0 m/s as it strikes a spring whose force constant, k = 20 N/m.

A. How far will the spring compress before it comes to rest ?

B. What is the speed of the mass when the compression of the spring is 0.20 m ?

given : m = 0.20 kg ; $v_o = 3.0$ m/s ; k = 20 N/m

(a) $U_1 + K_1 = U_2 + K_2$

$0 + (1/2)(0.20\text{ kg})(3.0\text{ m/s})^2 = (1/2)(20\text{ N/m})\, x^2 + 0$

$x = 0.30\text{ m}$

(b) $U_1 + K_1 = U_3 + K_3$

$0 + (1/2)(0.20\text{ kg})(3.0\text{ m/s})^2 = (1/2)(20\text{ N/m})(0.20\text{ m})^2 + (1/2)(0.20\text{ kg})v_3^2$

$v_3 = 2.2\text{ m/s}$

Example 4 A 0.20 kg mass is dropped 2.0 m above a vertical spring, which is initially uncompressed. The force constant of the spring is 20 N/m

A. What is the speed of the mass as it strikes the spring ?

B. How far will the mass compress the spring ? (consider gravitational potential energy)

given : m = 0.20 kg ; y = 2.0 m ; k = 20 N/m

(a) $U_1 + K_1 = U_2 + K_2$

$mg\Delta h + 0 = (1/2)mv^2$

$(0.20\text{ kg})(9.80\text{ m/s}^2)(2.0\text{ m}) = (1/2)(0.20\text{ kg})\, v^2$

$v = 6.26\text{ m/s}$

(b) $U_1 + K_1 = U_3 + K_3$

$(0.20\text{ kg})(9.80\text{ m/s}^2)(2.0\text{ m} + x) + 0 = (1/2)(20\text{ N/m})\, x^2 + 0$

$1.96\,(2.0 + x) = 10\, x^2$; $3.92 + 1.96\, x = 10\, x^2$

$0 = 10\, x^2 - 1.96\, x - 3.92$ (now use quadratic equation)

$x = 0.73\text{ m}$

5.6 Power

Example 1 How much work (in joules) will a 1.5 hp motor deliver in 5.0 min ?

given : P = 1.5 hp ; t = 5.0 min

$P = (1.5\ hp)\ (746\ W / hp) = 1.1 \times 10^3\ W$

$P = W / t$ $\quad$ $t = 5.0\ min\ (60\ s / min) = 3.0 \times 10^2\ s$

$(1.1 \times 10^3\ W) = W / (3.0 \times 10^2\ s)$; $W = 3.3 \times 10^5\ J$

Example 2 A 1.5-kg mass is lifted with a speed of 0.50 m/s. What power must a motor deliver to lift the mass ?

given : m = 1.5 kg ; v = 0.50 m/s

$P = F v$ $\quad$ $F = mg$; $F = (1.5\ kg)(9.80\ m/s^2) = 15\ N$

$P = (15\ N)(0.50\ m/s) = 7.5\ W$

Solutions to selected paired problems and selected problems

3. no work can be done if there is no motion, because without motion there is no displacement.

8. given : F = 250 N ; $W = 1.44 \times 10^3\ J$; $\theta = 30°$

   $W = F d (\cos \theta)$; $1.44 \times 10^3\ J = (250\ N)\ (d)\ (\cos 30°)$; $d = 6.65\ m$

14. first, the force the father exerts must be found ; since the velocity is constant the net force is zero.

    $0 = F \cos \theta - f_k$; $0 = N + F \sin \theta - mg$ (see example 6 in friction section in Ch.4)

    $\mu_k = F (\cos \theta) / (mg - F \sin \theta)$; $F = \mu_k mg / (\cos \theta + \mu_k \sin \theta)$

    $F = [(0.25)(35\ kg)(9.80\ m/s^2)] / [(\cos 30°) + (0.25)(\sin 30°)]$; $F = 87\ N$

    now the work can be found : $W = Fd \cos \theta$;

    $W = (87\ N)(10\ m)(\cos 30°) = 7.5 \times 10^2\ J$

22. first use Hooke's law : $F = kx$; $(0.25\text{ kg})(9.80\text{ m/s}^2) = k\,(0.050\text{ m})$
$k = 49\text{ N/m}$
work done by a spring : $W = (1/2)kx^2$; $6.0\text{ J} = (1/2)(49\text{ N/m})(x^2)$; $x = 0.49\text{ m}$

26. area under a F vs. x graph gives the amount of work done.
$W = (1/2)(2.0\text{ m})(6.0\text{ N}) + (1/2)(3.0\text{ m})(-6.0\text{ N}) = -3.0\text{ J}$

31. to find the one with the greatest kinetic energy substitute $K = (1/2)mv^2$
(a) $K = (1/2)(4\text{ m})(v)^2 = 2\text{ mv}^2$ (b) $K = (1/2)(3\text{ m})(2\text{ v})^2 = 6\text{ mv}^2$
(c) $K = (1/2)(2\text{ m})(3\text{ v})^2 = 9\text{ mv}^2$ (d) $K = (1/2)(\text{m})(4\text{ v})^2 = 8\text{ mv}^2$
the correct answer is (c)

36. the work done equals the change in kinetic energy
$W = \Delta K = K - K_o$; $W = 0 - (1/2)mv_o^2$
$W = Fd\cos\theta$; $F\,d\cos(180°) = -(1/2)mv_o^2$
since F and m are constants the stopping distance is proportional to the square of its instantaneous speed or veocity.

40. (a) the kinetic energy of the mass will be stored as potential energy in the spring - law of conservation of energy.
$K_1 + U_1 = U_2 + K_2$
$(1/2)(0.150\text{ kg})(4.75\text{ m/s})^2 + 0 = (1/2)(200\text{ N/m})\,x^2 + 0$; $x = 0.130\text{ m}$
(b) $K_1 + U_1 = U_3 + K_3$
$(1/2)(0.150\text{ kg})(4.75\text{ m/s})^2 + 0 = (1/2)(200\text{ N/m})(0.0600\text{ m})^2 + K_3$
$K = 1.33\text{ J}$; $K = (1/2)\,mv^2$; $1.33\text{ J} = (1/2)(0.150\text{ kg})\,v_3^2$; $v_3 = 4.21\text{ m/s}$
(c) when the mass recoils, the energy stored in the spring will be zero, therfore all of the energy is in the form of kinetic energy -- same as when the mass strikes the spring 4.75 m/s

46. (a) the zero height level is at the ground level.
in the basement : $U = mgh$; $U = (1.5\text{ kg})(9.80\text{ m/s}^2)(-8.0\text{ m}) = -44\text{ J}$
in the attic : ; $U = (1.5\text{ kg})(9.80\text{ m/s}^2)(4.5\text{ m}) = 66\text{ J}$
(b) attic to ground level would reduce the potential energy : - 66 J
basement to the attic would increase the potential energy: 44 J + 66 J =110 J

54. the speed is the greatest at the lowest point of its swing -- potential energy the least and kinetic energy the greatest. (c)

56. the total mechanical energy remains constant in this conservative system

(a) $U_1 + K_1 = U_2 + K_2$; U = mgh (directly proportional to height)

$0 + 80\ J = 60\ J + K_2$; $K_2 = 20\ J$; $U_2 = 60\ J$

(b) $K = (1/2)\ mv^2$; $20\ J = (1/2)(0.50\ kg)\ v^2$; v = 8.9 m/s

(c) at the maximum height the kinetic energy is zero , therefore the potential energy is 80 J . (Note : this is not the case for projectile motion since at the peak height, the mass is moving with a v_x.)

60. $U_1 + K_1 = U_2 + K_2 + Q$

$mgh_1 + 0 = 0 + (1/2)mv_2{}^2 + Q$

$(28\ kg)(9.80\ m/s^2)(3.0\ m) = (1/2)(28\ kg)(2.5\ m/s)^2 + Q$; $Q = 7.3 \times 10^2\ J$

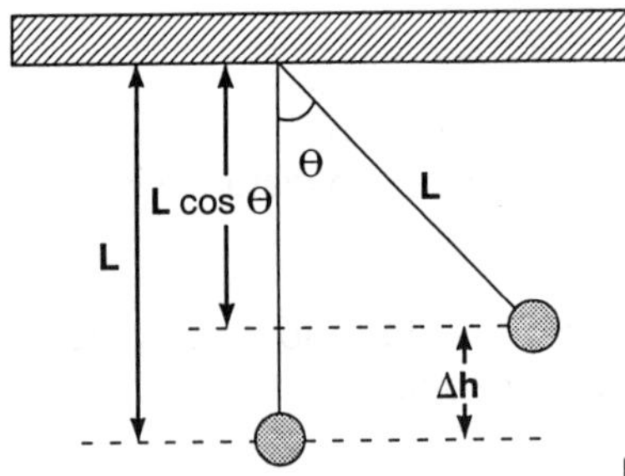

Fig. 5E.63

63. (a) $\Delta h = L - L\cos\theta$; $\Delta h = L\ (1 - \cos 25°)$

(b) $U_1 + K_1 = U_2 + K_2$

$\Delta h_2 = (0.75\ m)\ (1 - \cos 9.0°) = 0.0092\ m$

$(0.15\ kg)(9.80\ m/s^2)(0.070\ m) = (0.15\ kg)(9.80\ m/s^2)(0.0092\ m) + K_2$

$K_2 = 9.0 \times 10^{-2}\ J$

(c) $U_1 + K_1 = U_3 + K_3$;

$\Delta h_1 = L - L\cos\theta$; $\Delta h_1 = (0.75\ m) - (0.75\ m)(\cos 25°) = 0.070\ m$

$mgh_1 + 0 = 0 + (1/2)mv^2$

$(0.15\ kg)(9.80\ m/s^2)(0.070\ m) = (1/2)(0.15\ kg)(v^2)$; v = 1.4 m/s

73. (a) no ; power is a measure of the rate work is done
(b) if the efficiency is 100% there is no loss of energy
if the efficiency could be greater than 100 %, there would be more energy output, than input, which is a violation of the law of conservation of energy.

78. given : $P = 3.5$ hp (550 ft-lb/s / hp) $= 1.9 \times 10^3$ ft-lb/s ; $v = 1.5$ ft/s
(a) $P = Fv$; 1.9×10^3 ft-lb/s $= F$ (1.5 ft/s) ; $F = 1.3 \times 10^3$ lb
(b) since the net work is zero the rate at which mower is delivering power must equal the rate friction is doing work : -1.3×10^3 lb
(c) $P = W / t$; 1.9×10^3 ft-lb/s $= W / 6.0$ s ; $W = 1.1 \times 10^4$ ft-lb

84. first find the net work done : $W = \Delta U + \Delta K$

$$W = mg\Delta h + (1/2)mv^2$$

$$W = (3.25 \times 10^3 \text{ kg})(9.80 \text{ m/s}^2)(10.0 \times 10^3 \text{ m}) + (1/2)(3.25 \times 10^3)(236 \text{ m/s})^2$$

$$W = 4.09 \times 10^8 \text{ J}$$

$P = W / t$; $1.12 \times 10^6 = W / 750$ s   1500 hp $= 1.12 \times 10^6$ W

$W = 8.4 \times 10^8$ J

$\varepsilon = W_{out} / W_{in}$; $\varepsilon = (4.09 \times 10^8 \text{ J}) / (8.4 \times 10^8 \text{ J}) = 48.9$ %

Sample Quiz

Multiple Choice. Choose the correct answer.

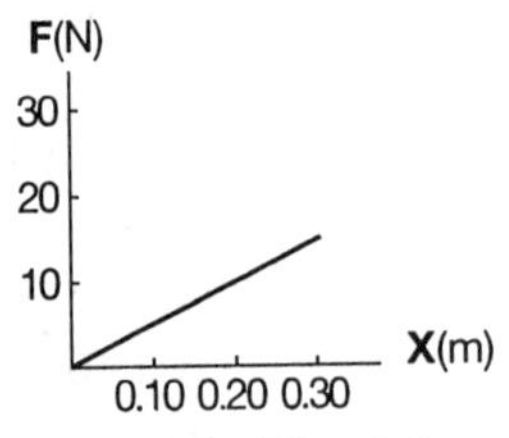

Fig. 5.2

Questions 1 and 2 refer to the graph in Fig. 5.2.

__ 1. What is the force constant ?
A. 25 N/m B. 50 N/m C. 100 N/m D. 150 N/m

__ 2. What is the work done as the object moves from 0 to 0.30 m ?
A. 2.3 J B. 4.5 J C. 9.0 J D. 14 J

__ 3. When a car is traveling with a speed of 55 mi/h it takes the driver a distance d to stop. If the driver is traveling at 70 mi/h, what distance in terms of d is needed to stop the car ?
A. d B. 1.3 d C. 1.6 d D. 2.6 d

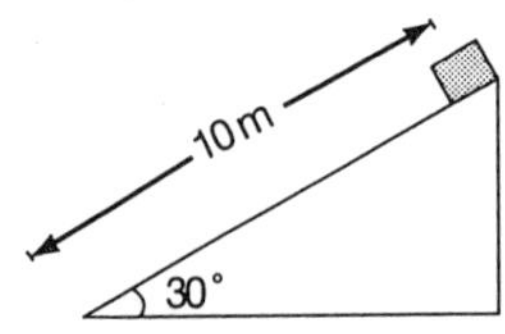

Fig. 5.3

__ 4. A mass, initially at rest, slides 10 m down a 30° frictionless incline as shown in Fig. 5.3. The speed of the mass at the bottom of the plane is
A. 10 m/s B. 14 m/s C. 100 m/s D. 200 m/s

__ 5. The work required a compress spring a distance x is 10 J. How much more work is required to compress the spring a distance 2x ?
A. 10 J B. 20 J C. 30 J D. 40 J

__ 6. A 10-kg mass is moving with a speed of 4.0 m/s. How much work is required to stop the mass ?
A.- 40 J B. -57 J C. -80 J D. -113 J

__ 7. A 10 N force is needed to move a mass with a constant speed of 5.0 m/s. What power must be delivered to the mass ?
A. 0.50 W B. 2.0 W C. 50 W D. 100 W

Problems

8. A force of 10 N is applied horizontally to a 2.0-kg mass on a horizontal surface. The coefficient of kinetic friction between the mass and the surface is 0.20. The mass is moved a distance of 10 m.
   A. Find the work done by the applied force.
   B. Find the work done by the frictional force.
   C. Find the net work done in moving the mass.
   D. Find the change in kinetic energy of the mass.

9. An 80-kg diver falls from a board into water 44 m below the water.
   A. Find the potential energy of the diver on the board.
   B. Find the kinetic energy of the diver as he strikes the water.
   C. Find the potential and kinetic energy of the diver after he has fallen 10 m.
   D. How long does it take the diver to strike the water ?

10. A 150-kg elevator starts from rest 10.0 s later is observed to be moving 5.0 m/s at a height of 50 m above the intitial point. Find the power of the elevator motor.

CHAPTER 6 Momentum and Collisions

Chapter Objectives

Upon completion of the unit on momentum, students should be able to :

1. define momentum and impulse.
2. write the expression that relates velocity, mass, and linear momentum, and recognize that momentum is a vector.
3. given a net force applied to a body over a period of time, find the change in momentum.
4. express Newton's second law of motion in terms of momentum.
5. relate the change in momentum to the area under a graph of force versus time.
6. state and apply the law of conservation of momentum.
7. identify situations in which linear momentum is conserved.
8. distinguish between internal and external forces.
9. apply the law of conservation of linear momentum for bodies in one or two dimensions, for elastic, inelastic, and totally inelastic collisions.
10. define the center of mass.
11. calculate the center of mass for a discrete group of particles.

Chapter Summary

* The **linear momentum** is an object is the product of its mass and velocity.

* Force may be expressed as the time rate of change of momentum or $\Delta p / \Delta t$, which is equivalent to the product of mass and acceleration if the mass is constant.

* **The law of conservation of linear momentum** says that the total linear momentum of a system is conserved if the net external force is constant.

* The **impulse-momentum theorem** states that the impulse is equal to the change in momentum.

* The total kinetic energy is conserved in an **elastic collision**. The total kinetic energy is not conserved in an **inelastic collision**. If the colliding objects stick together, the collision is a **completely inelastic collision**. For isolated systems, the momentum is conserved in both elastic and inelastic collisions.

* The **center of mass** is the point at which all the mass of an object or system may considered to be concentrated in representing the object of system as a particle. If the acceleration due to gravity is constant, the **center of gravity**, or the point at which all the weight of an object may be considered to be concentrated, is at the same location as the center of mass.

*For a system of particles, the x coordinate of the center of mass is found by summing the products of the masses of the particles and their distances from the origin and dividing by the sum of the total mass. In two dimensions, the location of the center of mass is (X_{cm}, Y_{cm}) where Y_{cm} is found using the same procedure as for X_{cm}.

Important Terms and Relationships

6.1 Linear Momentum

linear momentum : $\mathbf{p} = m\mathbf{v}$
total linear momentum of a system : $\mathbf{p} = \mathbf{p}_1 + \mathbf{p}_2 + \mathbf{p}_3 + \ldots$
Newton's second law in terms of momentum : $\mathbf{F} = \Delta\mathbf{p} / \Delta t$

6.2 The Conservation of Linear Momentum

conservation of linear momentum
(external vs. internal forces)

6.3 Impulse

impulse
impulse-momentum theorem : $\mathbf{F}\Delta t = \Delta\mathbf{p} = m\mathbf{v} - m\mathbf{v}_o$
area for F vs. t gives impulse or change in momentum

6.4 Elastic and Inelastic Collisions

elastic collision : (conditions : $K_f = K_i$ and $p_f = p_i$)
inelastic collision : (conditions : $K_f < K_i$ and $p_f = p_i$)
completely inelastic collision (conditions : $K_f < K_i$ and $p_f = p_i$)
final velocities in head-on two body collision (elastic collision)

$$v_1 = (m_1 - m_2)\, v_{1o} / (m_1 + m_2)$$

$$v_2 = (2\, m_1 v_{1o}) / (m_1 + m_2)$$

6.5 Center of Mass

center of mass (CM) : $X_{CM} = \Sigma_i (m_i x_i) / M$
center of gravity (CG)

6.6 Jet Propulsion and Rockets

jet propulsions
reverse thrust

Additional Solved Problems

6.1 Linear Momentum

Example 1 A 10-g bullet travels with a speed of 350 m/s. How fast must a 1750 kg car travel to have the same momentum as the bullet ?

first find the magnitude of the p for the bullet : $p = mv = (0.010 \text{ kg})(350 \text{ m/s}) = 3.5$ kg-m/s

now find the speed of the car :

$$p = mv \text{ ; } 3.5 \text{ kg-m/s} = (1750 \text{ kg})\, v \text{ ; } v = 2.0 \times 10^{-3} \text{ m/s}$$

Example 2 A 0.10 kg mass strikes a wall with a speed of 10 m/s and rebounds with a speed of 10 m/s. What is the change in momentum ?

there is a change in momemtum since the velocity -- not speed -- changes. taking v_o to be in the positive direction.

$$\Delta p = p - p_o \text{ ; } \Delta p = mv - mv_o = (0.10 \text{ kg})(-10 \text{ m/s}) - (0.10 \text{ kg})(10 \text{ m/s})$$

$$\Delta p = 2.0 \text{ kg-m/s in opposite direction of } v_o$$

6.2 The Law of Conservation of Momentum

Example 1 A 70-kg back running at 2.0 m/s runs into a 90-kg tackler moving at 1.0 m/s into the same direction. What is their common velocity immediately after they collide.

given : $m_1 = 70$ kg and $v_{1o} = 2.0$ m/s ; $m_2 = 90$ kg and $v_2 = -1.0$ m/s

$$p_o = p$$

$$m_1 v_{1o} + m_2 v_{2o} = (m_1 + m_2)\, v$$

$$(70 \text{ kg})(2.0 \text{ m/s}) + (90 \text{ kg})(-1.0 \text{ m/s}) = [(70 \text{ kg}) + (90 \text{ kg})]\, v$$

$v = 0.31$ m/s in the original direction of back

Example 2 Two persons are standing on ice. The 70-kg person pushes the 50-kg person giving the 50-kg person a speed of 1.5 m/s. What is the speed of the 70-kg person ?

given : $m_1 = 70$ kg ; $m_2 = 50$ kg ; $v_2 = 1.5$ m/s ; $v_{1o} = 0$; $v_{2o} = 0$

$$p_o = p$$

$$0 = m_1 v_1 + m_2 v_2$$

$$0 = (70 \text{ kg})(v_1) + (50 \text{ kg})(1.5 \text{ m/s})$$

$v_1 = 1.1$ m/s in opposite direction of the 50-kg person

Example 3 A 10-g bullet is fired into a 1.00-kg block. The bullet emerges with a speed of 150 m/s and the block moves with a speed of 1.5 m/s. What was the initial speed of the bullet ?

given : $m_1 = 0.010$ kg ; $m_2 = 1.00$ kg ; $v_1 = 150$ m/s ; $v_2 = 1.5$ m/s

$$p_o = p$$

$$m_1 v_{1o} = m_1 v_1 + m_2 v_2$$

$$(0.010 \text{ kg})(v_{1o}) = (0.010 \text{ kg})(150 \text{ m/s}) + (1.00 \text{ kg})(1.5 \text{ m/s})$$

$$v_{1o} = 300 \text{ m/s}$$

6.3 Impulse

Example 1 The impulse exerted on a 0.20-kg mass is 100 N-s.
A. What is the magnitude of the change in momentum ?
B. What is the magnitude of the change in velocity ?

given : $m = 0.20$ kg ; $\Delta p = 100$ N-s

A. the change in momentum equals the impulse imparted --- 100 N-s

B. $\Delta p = m\Delta v$; 100 N-s = (0.20 kg)Δv ; $\Delta v = 5.0 \times 10^2$ m/s

Example 2 A 0.20-kg ball is released from a height of 2.0 m , bounces on the floor and rebounds to a height of 1.7 m. The ball is in contact with the floor for 0.010 s.

A. Find the change in the ball's momentum when it strikes the floor.

B. Find the force the floor exerts on the ball.

given : $m = 0.20$ kg ; $h_1 = 2.0$ m ; $h_2 = 1.7$ m ; $t = 0.010$ s

(a) using energy conservation : $mgh_1 = (1/2)mv_1^2$;

$$(9.80\ \text{m/s}^2)(2.0\ \text{m}) = (1/2)\ v_1^2\ ;\ v_1 = -6.3\ \text{m/s}$$

$$(1/2)mv_2^2 = mgh_2\ ;\ (1/2)\ v_2^2 = (9.80\ \text{m/s}^2)(1.7\ \text{m})\ ;\ v_2 = 5.8\ \text{m/s}$$

$$\Delta p = mv - mv_o = (0.20\ \text{kg})(5.8\ \text{m/s}) - (0.20\ \text{kg})(-6.3\ \text{m/s}) = 2.4\ \text{kg-m/s}$$

(b) $F = m\Delta v / t = (0.20\ \text{kg})(12.1\ \text{m/s}) / (0.010\ \text{s}) = 2.4 \times 10^2$ N

6.4 Elastic and Inelastic Collisions

Example 1 A 2.0-kg mass moves in the -x direction with a speed of 4.0 m/s when it collides with a 3.0-kg mass moving with a speed of 2.0 m/s in the +y direction. If the masses undergo a total inelastic collision, find the velocity immediately after the collision occurs.

given : $m_1 = 2.0$ kg ; $v_{1ox} = -4.0$ m/s ; $m_2 = 3.0$ kg ; $v_{2oy} = 2.0$ m/s

x : $m_1 v_{1o} = (m_1 + m_2)\ v\ (\cos\theta)$

$(2.0\ \text{kg})(-4.0\ \text{m/s}) = (5.0\ \text{kg})\ v\ (\cos\theta)$

$1.6 = v\ (\cos\theta)$

y : $m_2 v_{2o} = (m_1 + m_2)\ v\ (\sin\theta)$

$(3.0\ \text{kg})(2.0\ \text{m/s}) = (5.0\ \text{kg})\ v\ (\sin\theta)$

$1.2 = v\ (\sin\theta)$

to solve divide the y equation by the x equation

$$v\ (\sin\theta)\ /\ v\ (\cos\theta) = 1.2 / 1.6\ ;\ \tan\theta = 0.75\ ;\ \theta = 37^\circ \text{ above -x axis}$$

to find v substitute into either x or y equation: $1.6 = v\ (\cos 37^\circ)$; $v = 2.0$ m/s

Example 2 A 2.0-kg mass slides down a 30° frictionless inclined plane 10 m long. At the bottom of the plane the mass slides on a horizontal frictionless surface until it collides elastically with a stationary 5.0-kg mass. Find the velocity of each mass immediately after the collision.

given : $m_1 = 2.0$ kg ; $m_2 = 5.0$ kg ; d = 10 m therefore y = 5.0 m

first use the law of conservation of energy to find the speed of the 2.0-kg mass before the collision.

$$mg\Delta h = (1/2)mv_o^2 \; ; \; (2.0 \text{ kg})(9.80 \text{ m/s}^2)(5.0 \text{ m}) = (1/2)(2.0 \text{ kg})\, v_o^2$$

$$v_o = 9.9 \text{ m/s}$$

$v_2 = 2\, m_1 v_{1o} / (m_1 + m_2)$

$v_2 = 2\,(2.0 \text{ kg})(9.9 \text{ m/s}) / [\,(2.0 \text{ kg}) + (5.0 \text{ kg})] = 5.6 \text{ m/s}$

now using the law of conservation of momentum :

$$m_1 v_{1o} = m_1 v_1 + m_2 v_2$$

$$(2.0 \text{ kg})(9.9 \text{ m/s}) = (2.0 \text{ kg})\, v_1 + (5.0 \text{ kg})(5.6 \text{ m/s})$$

$$v_1 = -4.1 \text{ m/s}$$

after the collision $v_1 = 5.6$ m/s in the original direction of v_{1o}

and $v_2 = 4.1$ m/s in the opposite direction of v_{1o}

6.5 Center of Mass

Example 1 Find the location of the center of mass of the four mass system.

mass	location
$m_1 = 4.0$ kg	(0 , -5.0 m)
$m_2 = 3.0$ kg	(- 2.0 m , 2.0 m)
$m_3 = 1.0$ kg	(2.0 m , 0)
$m_4 = 6.0$ kg	(0 , 5.0 m)

$\Sigma\, X_{CM} = (\, m_1x_1 + m_2x_2 + m_3x_3 + m_4x_4) / M$

$X_{CM} = [\, (4.0 \text{ kg})(0) + (3.0 \text{ kg})(-2.0 \text{ m}) + (1.0 \text{ kg})(2.0 \text{ m}) + (6.0 \text{ kg})(0)\,] / 14 \text{ kg}$

$X_{CM} = -0.28$ m

$\Sigma\, Y_{CM} = (m_1y_1 + m_2y_2 + m_3y_3 + m_4y_4) / M$

$Y_{CM} = [(4.0 \text{ kg})(-5.0 \text{ m}) + (3.0 \text{ kg})(2.0 \text{ m}) + (1.0 \text{ kg})(0) + (6.0 \text{ kg})(5.0 \text{ m})] / 14 \text{ kg}$

$y_{CM} = 1.14$ m

$(X_{Cm}, Y_{Cm}) = (-0.28 \text{ m} , 1.14 \text{ m})$

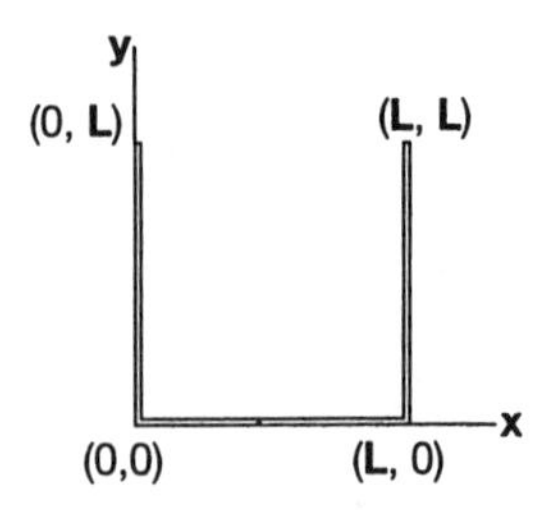

Fig. 6.1

Example 2 In Fig. 6.1, the mass is uniformly distributed in the three rods of mass m. Locate the center of mass of the system in terms of L.

let m_1 be on left , m_2 on the bottom, and m_3 on the right

$$\Sigma X_{CM} = (m_1x_1 + m_2x_2 + m_3x_3) / M$$
$$= [(m)(0) + (m)(L / 2) + (m) (L)] / (3 m) = 0.50 L$$

$$\Sigma Y_{CM} = (m_1y_1 + m_2y_2 + m_3y_3) / M$$
$$= [(m)(L/2) + 0 + m(L/2)] / (3 m) = 0.33 L$$

(X , Y) = (0.50 L , 0.33 L)

Solutions to paired problems and selected problems.

5. $p = mv$; $p = (7.3 \text{ kg})(20 \text{ m/s}) = 1.5 \times 10^2$ kg-m/s

10. $p = p_1 - p_2$ (subtract -- since the electrons are moving in opposite directions)

$p = m_1v_1 - m_2 v_2$;

$p = (9.11 \times 10^{-31}\text{kg})(4.5 \times 10^2 \text{ m/s}) - (9.11 \times 10^{-31}\text{kg})(3.4 \times 10^2 \text{ m/s})$

$p = 1.0 \times 10^{-28}$ kg-m/s in the direction of the faster electron

16. take into consideration the change in momentum of both components. since the x - velocity does not change , there is no change in linear momentum in this component ; yet, there is a change in momentum in the y since the perticle is traveling in the opposite direction.

$\Delta p = mv \cos 60° - (-mv \cos 60°)$; $\Delta p = 2 mv (\cos 60°)$

$\Delta p = 2 (0.20 \text{ kg})(15 \text{ m/s})(\cos 60°) = 3.0$ kg-m/s

21. (a) they cancel each other out according to Newton's third law of motion

26. let m_1 and v_1 represent the bullet and m_2 and v_2 represent the block

$p_o = p$

$m_1 v_{1o} = (m_1 + m_2)\,v$ (v is the same for both bullet and block)

(0.100 kg)(250 m/s) = (0.100 kg + 14.9 kg) v

v = 1.67 m/s in the direction of the bullet

28. treat x and y componets separately

$m = m_1 + m_2$; let $m_1 = 2.0$ kg , $m_2 = 1.0$ kg

y : $m_o v_o = m_1 v_{1y} + m_2 v_2$

(3.0 kg)(2.0 m/s) = (2.0 kg)(v_{1y}) + 0

$v_{1y} = 3.0$ m/s

x : $m_o v_o = m_1 v_{1x} + m_2 v_2$

0 = (2.0 kg)(v_{1x}) + (1.0 kg)(3.7 m/s)

$v_{1x} = -1.9$ m/s

now find the resultant velocity of mass 1 : $v^2 = v_x{}^2 + v_y{}^2$

$v^2 = (-1.9 \text{ m/s})^2 + (3.0 \text{ m/s})^2$

v = 3.6 m/s

to find the direction : tan θ = 3.0 / 1.9 ; θ = 58° above the -x axis (2nd quadrant)

34. $p_o = p$

$m_1 v_1 = (m_1 + m_2)\,v$

(1600 kg)(2.5 m/s) = (1600 kg + 3500 kg) v

v = 0.78 m/s in the original direction of the car

40. (a) if the masses are equal the balls interchange velocities thus they interchange kinetic energies since their masses are equal.

50. (a) impules equals the change in momentum :

$\Delta p = m\Delta v$; $\Delta p = (0.15 \text{ kg})(0 - 8.0 \text{ m/s}) = -1.2$ N-s ($\Delta v = v - v_o$)

(b) $F = \Delta p / \Delta t$; F = (-1.2 N-s) / (0.10 s) = -12 N (negative signifies force is in the opposite direction of v_o)

54. (a) note the key word <u>elastic</u>

$v_2 = 2\, m_1 v_{1o} / (m_1 + m_2)$

$v_2 = 2\,(1.67 \times 10^{-27} \text{ kg})(4.00 \times 10^5 \text{ m/s}) / [\,(1.67 \times 10^{-27} \text{ kg}) + (6.64 \times 10^{-27} \text{ kg})]$

$v_2 = 1.61 \times 10^5$ m/s

go back to conservation of momentum

$m_1 v_{1o} = m_1 v_1 + m_2 v_2$

$(1.67 \times 10^{-27})(4.00 \times 10^5) = (1.67 \times 10^{-27})\, v_1 + (6.64 \times 10^{-27})(1.61 \times 10^5)$

$v_1 = -2.39 \times 10^5$ m/s

(b) $K_{1o} = (1/2)\, m_1 v_{1o}^2 = (1/2)(1.67 \times 10^{-27}\text{ kg})(4.00 \times 10^5\text{ m/s})^2 = 1.33 \times 10^{-16}$ J

$K_2 = (1/2)\, m_2 v_2^2 = (1/2)(6.64 \times 10^{-27}\text{ kg})(1.61 \times 10^5\text{ m/s})^2 = 8.61 \times 10^{-17}$ J

$K_2 / K_{1o} = 64.4$ %

58. first find the speed of the wad when it strikes the surface ; (free fall of conservation of energy)

$mgh = (1/2)\, mv^2$; $(0.35\text{ kg})(9.80\text{ m/s}^2)(1.5\text{ m}) = (1/2)(0.35\text{ kg})\, v^2$

velcoty of wad when strikes the ground is -5.4 m/s

$F = \Delta p / \Delta t$; $F = m\Delta v / \Delta t$; $F = (0.35\text{ kg})[0 - (-5.4\text{ m/s}] / (0.10\text{ s}) =$ 19 N upward

62. x : $mv = (m + M)\, v_x'$

y : $MV = (m + M)\, v_y'$

$v_x' = mv / (m + M)$; $v_y' = MV / (m + M)$

66. first find the velocity of the car before it collides :

$mgh = (1/2)\, mv_o^2$

$(2.0 \times 10^4\text{ kg})(9.80\text{ m/s}^2)(3.3\text{ m}) = (1/2)(2.0 \times 10^4\text{ kg})\, v_o^2$; $v_o = 8.0$ m/s

now find the velocity of the combination after the collision

$m_1 v_{1o} = (m_1 + m_2)\, v$; $v = 4.0$ m/s

$K_o = (1/2)\, m_1 v_{1o}^2 = (1/2)(2.0 \times 10^4\text{ kg})(8.0\text{ m/s})^2 = 6.4 \times 10^5$ J

$K = (1/2)(2\, m_1)\, v^2 = (1/2)(4.0 \times 10^4\text{ kg})(4.0\text{ m/s})^2 = 3.2 \times 10^5$ J

$K_o / K = 50$ %

70. momentum is conserved : $m_1v_{1o} = (m + M)\,v$

solve for v : $v = mv_o / (m + M)$

$K_{lost} / K_o = 1 - [\,m / (m + M)\,]$

$K_{lost} / K_o = (m + M - m) / (m + M) = M / (m + M)$

82. by inspection all masses have a zero y coordinate -- therefore $Y_{CM} = 0$

$X_{CM} = (m_1x_1 + m_2x_2 + m_3x_3) / M$

$Y_{CM} = [\,(3.0\text{ kg})(-6.0\text{ m}) + (2.0\text{ kg})(1.0\text{ m}) + (4.0\text{ kg})(3.0\text{ m})\,] / (9.0\text{ kg}) = -0.44\text{ m}$

$(X, Y) = (-0.44\text{ m}, 0)$

91. notice the key word <u>elastic</u> .

$v_2 = 2\,m_1v_{1o} / (m_1 + m_2)$

$v_2 = [\,2\,(2.0\text{ kg})(6.0\text{ m/s})\,] / [(2.0\text{ kg}) + (10\text{ kg})] = 2.0\text{ m/s}$

now using law of conservation of momentum

$m_1v_{1o} = m_1v_1 + m_2v_2$

$(2.0\text{ kg})(6.0\text{ m/s}) = (2.0\text{ kg})(\,v_1\,) + (10\text{ kg})\,(2.0\text{ m/s})$; $v_1 = -4.0\text{ m/s}$

the 10-kg mass moves forward with a velocity of 2.0 m/s and the 2.0-kg mass recoils with a speed of 4.0 m/s.

102. treat x and y components independently :
momentum is conserved

x: $mv_o = m_1v_1 + m_2v_2 + m_3\,v_3$

$0 = (0.50\text{ kg})(-2.8\text{ m/s}) + 0 + (1.2\text{ kg})(\,v_3)(\cos\theta)$; $v_3\cos\theta = 1.2$

y: $mv_o = m_1v_1 + m_2v_2 + m_3\,v_3$

$0 = 0 + (1.3\text{ kg})(-1.5\text{ m/s}) + (1.2\text{ kg})\,v_3\,(\sin\theta)$; $v_3\sin\theta = 1.6$

now take the equations and divide : $v_3\sin\theta / v_3\cos\theta = 1.6 / 1.2$

$\tan\theta = 1.33$; $\theta = 53°$

to find v_3 substitute back into either equation. $v_3\,(\cos 53°) = 1.2$; $v_3 = 2.0\text{ m/s}$

Sample Quiz

Multiple Choice. Choose the correct answer.

__ 1. A 10 N force acts on a 2.0-kg mass for 5.0 s. What is the change in momentum?
A. 20 kg-m/s B. 10 kg-m/s C. 25 kg-m/s D. 50 kg-m/s

__ 2. A 2.0-kg ball moves to the right with a speed of 2.0 m/s and strikes a 1.0-kg ball initially at rest. If the balls stick, their common velocity will be
A. 0.67 m/s B. 1.0 m/s C. 1.3 m/s D. 1.7 m/s

__ 3. A 3.0-kg block slides at 2.0 m/s in the +x direction when it strikes a 2.0-kg mass moving in the -x direction with a speed of 3.0 m/s. After they collide, the masses stick. What is their common velocity ?
A. 1.0 m/s right B. 1.0 m/s left C. 2.0 m/s right D. 0

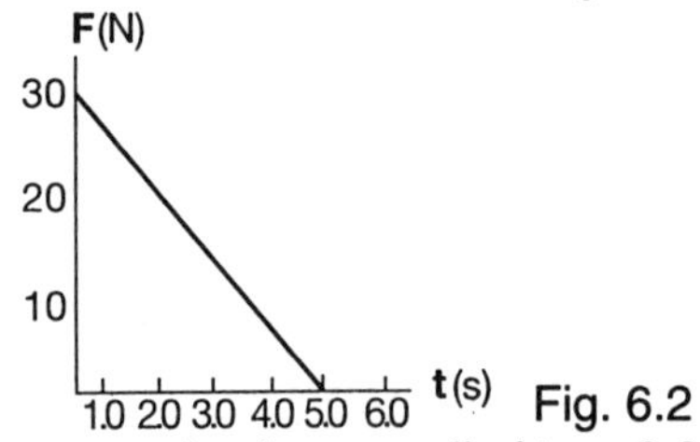

Fig. 6.2

__ 4. The graph in Fig. 6.2 represents the force applied to a 2.0-kg mass as a function of time. If the object is initially at rest, what is its speed after 5.0 s ?
A. 150 m/s B. 75 m/s C. 38 m/s D. 19 m/s

__ 5. The center of mass of two particles is at the origin. One particle is located at (3.0 m , 0 m) and has a mass of 2.0 kg. What is the location of 3.0-kg mass ?
A. -3.0 m B. -2.0 m C. 2.0 m D. 3.0 m

__ 6. A rider on an air track is traveling with a speed v when it makes an elastic collision with an identical rider. What is the speed of the rider initially in motion immediatley after the collision ?
A. - v B. 0 C. v D. cannot be determined

__ 7. A 20-g bullet is fired into a 9.98-kg block. The bullet becomes embedded in the block and together they travel with a speed of 2.00 m/s. What was the initial speed of the bullet ?
A. 3.0 m/s B. 20 m/s C. 5.0×10^2 m/s D. 1.0×10^3 m/s

Problems

8. A 0.200-kg mass is attached of a 0.800 m length string to form a pendulum. The pendulum is released from a horizontal position. At the bottom of its swing, it collides elastically with a 0.400-kg mass which is at rest on a horizontal, frictionless surface. Find the velocity of each mass immidately after the collision.

9. A 10-g bullet is fired into a 1.49-kg block. After the bullet becomes embedded in the block they move on a horizontal frictionless surface and eventually strike a spring whose force constant is 100 N/m. The spring is observed to undergo a maximum compression of 10 cm. From this information determine the initial speed of the bullet.

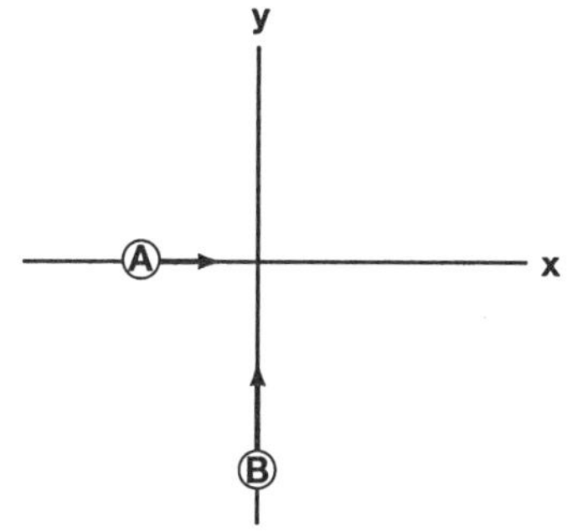

Fig. 6.3

10. In Fig. 6.3, the two masses are on a collision course. Particle A has a mass m and a velocity 2v directed in the +x direction. Particle B has a mass 2 m and a velocity 3v in the +y direction. The particles collide at the origin and become interlocked. Find their common velocity after the collision. (let v = 0.62 m/s)

CHAPTER 7 Circular Motion and Gravitation

Chapter Objectives

Upon completion of the unit on circular motion and gravitation, students should be able to :

1. use polar coordinates to write the position of a particle on a circular path.
2. convert units of angular distance to radians, degrees, or revolutions.
3. write expressions for angular kinematics to describe constant angular velocity, constant angular acceleration, and uniform angular acceleration.
4. apply the right-hand rule to determine the direction of angular velocity and acceleration.
5. distinguish between angular frequency and frequency.
6. state and write the relationships between period and frequency.
7. write expressions for tangential and angular quantities of motion.
8. calculate and give the direction for centripetal acceleration and force.
9. explain why an object can have a constant speed and yet not have constant velocity.
10. solve problems associated with centripetal acceleration and centripetal force.
11. state and write Newton's universal law of gravitation.
12. recognize and apply an inverse square law.
13. using Newton's law of gravitation, find the acceleration due to gravity at locations above the Earth's surface and for other astronomical bodies.
14. write an expression for the gravitational field of an object.
15. write an expression for the gravitational potential energy for a body in a nonuniform gravitational field.
16. apply the law of conservation of mechanical energy for bodies in a gravitational field.
17. define and calculate the escape speed for a body.
18. state and apply Kepler's three laws of motion.

Chapter Summary

*Angular (circular) motion is conveniently described using the polar coordinates (r, θ), because r is a constant and only θ varies.

*__Angular distance__ ($\Delta\theta$) is measured in degrees or radians. A **radian** (rad) is defined as the angle that subtends an arc length that is equal to the radius : 1 rad = 57.3° or 2π rad = 360°. In general, arc length (s) is equal to the product of the radius and the angle (in radians) subtended by the arc.

*__Angular speed__ and **angular velocity** have units of rad/s and are given by the equations analogous to those for linear motion. The direction of the angular velocity of an object in circular motion is given by a right-hand rule : if the fingers of the right hand are curled in the direction of the circular motion, the extended thumb points in the direction of the angular velocity. There are only two directions possible for a particular circular path, and they may be represented using a plus or minus sign.

*The **frequency** of an object in uniform circular motion is related to the **period** by $f = 1 / T$, and has the unit of hertz (Hz).

*For **uniform circular motion**, the speed is constant, but there is an acceleration (because of the change in direction of the velocity vector). This acceleration, called the **centripetal acceleration** (center seeking acceleration), is directed towards the center of the circular path and has a magnitude v^2 / r.

*Newton's second law of motion also applies to specific forces, i.e., weight ($w = mg$) or to motional descriptions such as centripetal force ($F = mv^2 / r$).

*__Angular acceleration__ has the units of rad/s^2 and is zero for uniform circular motion. The total acceleration for an object in nonuniform circular motion is given by $\mathbf{a} = a_t \mathbf{t} + a_c \mathbf{r}$, where a_t and a_c are the magnitudes of the instantaneous, tangential and centripetal acceleration, respectively.

*The particles of a uniformly rotating object has the same instantaneous angular speed and acceleration but different tangential speeds and acceleration at different radii.

*The **universal law of gravitation** states that the mutual gravitational attraction between any two particles is directly proportional to the product of their masses and inversely proportional to the square of the distance between their center of masses, ($1/r^2$) . The magnitude of the gravitational force is given by $F = Gm_1m_2/r^2$, where G is the **universal gravitational constant** and has a value of 6.67×10^{-11} N-m^2/kg^2.

*The acceleration due to the gravitational attraction on an objects by a mass M is a function of the distance between them (r); that is, $g(r) = GM/r^2$. Note the acceleration does not depend upon the mass of the object itself. At the surface of the Earth, $r = GM_e/R_e^2$. At an altitude h above the surface of the Earth, $r = R_e + h$, and $g(r) = GM_e / (R_e + h)^2$.

*The gravitational force per unit mass is a vector quantity called the gravitational intensity. which is used to map the gravitational field around a large object such as the Earth.

*The **gravitational potential energy** of two point masses is given by $U = -Gm_1m_2 / r$ and the total mechanical energy of a mass moving in the proximity of another mass is the sum of the kinetic energy of the moving mass and the gravitational potential energy of the two masses.

***Kepler's laws of planetary motion** are as follows :

1. The law of orbits Planets move in elliptical orbits with the Sun at one of the focal lengths.
2. The law of areas A line joining the Sun and a planet sweeps out equal areas in equal lengths of time.
3. The law of periods The square of the period of a planet is directly proportional to the cube of its average distance from the Sun. That is $T^2 = K r^3$, where $K = 2.97 \times 10^{-19}\ s^2/m$.

*The **escape speed** is the speed that must be given initially to an object for it to be raised to the top of a gravitational potential energy well of a planet or celestial body.

Important terms and Relationships

7.1 Angular Measurements

angular distance
radian (rad)
$s = r\theta$
1 revolution = 360° = 2π radians

7.2 Angular Speed and Velocity

average angular speed : $\omega_{av} = \Delta\theta / \Delta t$

constant angular speed : $\theta = \omega t$

angular velocity
average
instantaneous
tangential speed : $v = \omega r$
period
frequency : $f = 1 / T$
hertz (Hz)
angular speed with uniform circular motion : $\omega = 2\pi f$

7.3 Uniform Circular Motion and Centripetal Acceleration

uniform circular motion
centripetal acceleration : $a_c = v^2 / r = \omega^2 r$
centripetal force : $F = ma_c = mv^2 / r$

7.4 Angular Acceleration

average angular acceleration $\alpha_{av} = \Delta\omega / \Delta t$
constant acceleration $\omega = \omega_o + \alpha t$

$\theta = \omega_o t + (1/2)\alpha t^2$

$\omega^2 = \omega_o{}^2 + 2\alpha\theta$

tangential acceleration $a_t = \alpha r$
total acceleration $\mathbf{a} = \mathbf{a_t} + \mathbf{a_r}$

7.5 Newton's Law of Gravitation

universal law of gravitation : $F = G\, m_1 m_2 / r^2$
universal gravitational constant : 6.67×10^{-11} N-m^2 /kg^2
acceleration due to gravity : $g = G\, m / r^2$
g's of force
gravitational potential energy : $U = -\, Gm_1 m_2 / r$

7.6 Kepler's Laws and Earth Satellites

Kepler's first law (the law of orbits)
Kepler's second law (the law of areas)
Kepler's third law (the law of periods) : $T^2 = Kr^3$
escape speed : $v = [\, 2G\,M / r\,]^{1/2}$
energy of orbiting satellite : $E = -\,GmM / 2r$

Additional Solved Problems

7.1 Angular Motion

Example A car travels around a semicircle with a radius is 100 m.
A. What linear distance does the car travel ?
B. What angular distance does the car travel in radians and in degrees.

(a) $s = \pi\, r$; $s = (3.14)(100 \text{ m}) = 314 \text{ m}$
(b) $s = r\,\theta$; $314 \text{ m} = (100 \text{ m})\,\theta$; $\theta = 3.14 \text{ rad}$;
$(3.14 \text{ rad})\,(360° / 2\pi \text{ rad}) = 180°$

7.2 Angular Speed and Velocity

Example 1 A turntable is rotating at a rate of 33.3 rpm. How fast is a point 20.0 cm from the center of the turntable moving ?

given : $r = 100$ m ; $\theta = 2\pi$ rad $= 360°$

first the angular speed needs to be converted into rad/s

$(33.3 \text{ rev / min})\,(2\,\pi \text{ rad / rev})(\,1 \text{ min} / 60 \text{ s}) = 3.49 \text{ rad/s}$
$v = \omega\, r$; $v = (3.49 \text{ rad/s})(0.200 \text{ m}) = 0.697 \text{ m/s}$

Example 2 The moon takes 29.5 days to make a complete revolution around the Earth.
A. What is the anglar speed of the moon ?
B. What is the linear speed of the moon ?

given : $T = 29.5$ days ; $\theta = 1.0$ rev $= 2\pi$ rad

(a) $t = (29.5 \text{ days})(24 \text{ h / d}) (3600 \text{ s / h}) = 2.54 \times 10^6 \text{ s}$
$\omega = \theta / t$; $\omega = 2 \pi \text{ rad} / (2.54 \times 10^6 \text{ s}) = 2.46 \times 10^{-6} \text{ rad/s}$
(b) $v = \omega r$; $v = (2.46 \times 10^{-6} \text{ rad/s})(3.8 \times 10^8 \text{ m}) = 936 \text{ m/s}$

7.3 Uniform Circular Motion and Centripetal Acceleration

Example 1 A 0.50-kg mass moves in a circle whose radius is 5.0 m. It is noticed that it takes 2.0 s for the mass to move in a circle.
A. Find the speed of the mass.
B. Find the centripetal acceleration of the mass.

given : $m = 0.50$ kg ; $r = 5.0$ m ; $t = 2.0$ s

A. $v = 2\pi r / t$; $v = 2\pi(5.0 \text{ m}) / (2.0 \text{ s}) = 16 \text{ m/s}$
B. $a_c = v^2 / r$; $a_c = (16 \text{ m/s})^2 / (5.0 \text{ m}) = 51 \text{ m/s}^2$

Example 2 The coefficient of friction between the tires and the road is 0.80. How fast can a car round a horizontal curve if the radius of curvature is 80 m ?

given : $\mu = 0.80$; $r = 80$ m

the force which supplies the centripetal fore is the frictional force.

$ma = f$
$mv^2 / r = \mu N$ (horizontal surface : frictional force equals normal force)
$mv^2 / r = \mu mg$
$v^2 / (80 \text{ m}) = (0.80)(9.80 \text{ m/s}^2)$; $v = 25 \text{ m/s}$

Example 3 A 0.20-kg mass is attached to a 1.0 m long string to form a pendulum. The pendulum is released from a rest when the angle with the horizontal is 53°.
A. What is the speed of the pendulum at the bottom of the circle ?
B. What is the tension in the string at the bottom of the circle ?

given : $m = 0.50$ kg ; $L = 1.0$ m ; $\theta = 53°$

(a) law of conservation of energy revisited :

$$U_1 + K_1 = U_2 + K_2$$

$$mgh + 0 = 0 + (1/2)\, mv^2$$

$$gL\,(1 - \cos\theta) = (1/2)\, v^2 \;\; ; \;\; (9.80 \text{ m/s}^2)(1.0 \text{ m})(1 - \cos 37°) = (1/2)\, v_2^{\,2}$$

$$v_2 = 2.0 \text{ m/s}$$

(b) $mv^2 / r = T - mg$

$$(0.20 \text{ kg})(2.0 \text{ m/s})^2 / (1.0 \text{ m}) = T - (0.20 \text{ kg})(9.80 \text{ m/s}^2) \;\; ; \;\; T = 2.8 \text{ N}$$

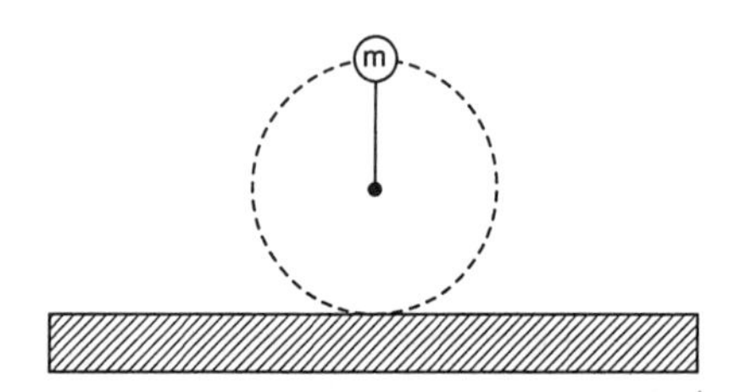

Example 4 A mass m moves in a vertical circle as shown above. The length of the string is L. At the top of the circle the tension in the string is equal to the object's weight.

A. Find the magnitude and the direction of the net force acting on the mass at the top of the circle.

B. Find the tension in the string when the mass is at the bottom of the circle.

C. Suppose the string breaks at the top of the circle.

(1) find the time it take for the mass to strike the ground.

(2) find the horizontal distance the mass travels before it strikes the ground.

given : mass m , length, L , and tension at top mg

A. $ma = T + mg$ (since the weight and the tension act in the same direction)

$ma = mg + mg = 2mg$ directed toward the center of the circle.

B. first use the law of conservation of energy to find the speed of the mass at the bottom of the circle ; also find the speed of the mass at the top of the circle.

$mv^2 / r = 2mg$; $v^2 = 2gL$

$U_1 + K_1 = U_2 + K_2$

$mg\,(2\,L) + (1/2)\,mv_1{}^2 = 0 + (1/2)\,mv_2{}^2$

$2\,mgL + (1/2)m\,(2gL) = (1/2)\,mv_2{}^2$

$6\,gL = v_2{}^2$;

$mv^2 / r = T - mg$; $mv_2{}^2 / L = T - mg$; $m(6\,gL) / L = T - mg$; $T = 7\,(mg)$

C. first find the time the mass is in the air : $y = (1/2)gt^2$

$2L = (1/2)gt^2$; $t = [2L / g]^{1/2}$

then find the horizontal distance : $x = v_x t$

$x = [2gL]^{1/2}[2\,L / g]^{1/2}$

$x = 2\,L$

7.4 Angular Acceleration

Example 1 A rotating disk starts from rest and is accelerated to 200 rpm in 30.0 s.
A. Find the angular acceleration of the mass.
B. Find the number of revolutions made by the disk during the 30 s interval.

given : $\omega_o = 0$; $\omega = 200$ rev/min = 20.9 rad/s ; t = 30.0 s

A. $\omega = \omega_o + \alpha t$; 20.9 rad /s = α (30.0 s) ; $\alpha =$ 0.70 rad/s^2

B. $\theta = (1/2)\alpha t^2$; $\theta = (1/2)(0.70\text{ rad/s}^2)(30.0\text{ s})^2 = 3.2 \times 10^2$ rad ;
$(3.2 \times 10^2$ rad) (1 rev / 2 π rad) =51 rev

Example 2 An object with uniform acceleration has an angular velocity of 50 rpm and a radius of 0.50 m. The object then accelerates at 0.50 rad/s^2 for a total of 30 s.
A. Find the angular speed after the inverval of acceleration.
B. Find the angular displacement during this time.
C. Find the total acceleration of the mass after the 30 s inverval.

given : $\omega_o = 50$ rev/min. ; $\alpha = 0.50$ rad/s^2 ; t = 30 s

A. $\omega = \omega_o + \alpha t$; ω_o = (50 rev/min)(2π rad / rev)(1 min / 60 s) = 5.2 rad/s

ω = (5.2 rad/s) + (0.50 rad/s^2)(30 s) = 20.2 rad/s

B. $\theta = (5.2 \text{ rad/s})(30 \text{ s}) + (1/2)(0.50 \text{ rad/s}^2)(30 \text{ s})^2 = 381 \text{ rad}$

C. $a_t = \alpha r$; $a_t = (0.50 \text{ rad/s}^2)(0.50 \text{ m}) = 0.25 \text{ m/s}^2$

$a_r = \omega^2 r = (5.2 \text{ rad/s})^2(0.50 \text{ m}) = 13.5 \text{ m/s}^2$

$\mathbf{a} = (0.25 \text{ m/s}^2)\,\mathbf{t} + (13.5 \text{ m/s}^2)\,\mathbf{r}$

7.5 Newton's Law of Gravitation

Example 1 What is the gravitational force of attraction between two 10-kg masses located a distance of 2.0 m from each other ?

given : $G = 6.67 \times 10^{-11}$ N-m^2 / kg^2 ; $m_1 = 10$ kg ; $m_2 = 10$ kg ; $r = 2.0$ m

$F = G\, m_1 m_2 / r^2$

$F = (6.67 \times 10^{-11} \text{ N-m}^2/\text{kg}^2)(10 \text{ kg})(10 \text{ kg}) / (2.0 \text{ m})^2 = 1.7 \times 10^{-9} \text{ N}$

Example 2 A hypothetical planet has a mass ten times that of Earth and a radius of half the Earth's. What is the acceleration due to gravity on the planet ?

given : $G = 6.67 \times 10^{-11}$ N-m^2 / kg^2 ; $m = 6.0 \times 10^{25}$ kg ; $r = 3.2 \times 10^6$ m

$g = G\, M / r^2$;

$g = (6.67 \times 10^{-11} \text{ N-m}^2/\text{kg}^2)(6.0 \times 10^{25} \text{ kg}) / (3.2 \times 10^6 \text{ m})^2$

$g = 3.9 \times 10^2 \text{ m/s}^2$

Example 3 Two 2.0-g point masses are separated by a distance of 1.0 m. One mass is held fixed and the other mass moves 40 cm. What is the speed of the mass at this location ?

given : $m_1 = m_2 = 2.0 \times 10^{-3}$ kg ; $r_1 = 1.0$ m ; $r_2 = 0.60$ m ;

$G = 6.67 \times 10^{-11}$ N-m^2 / kg^2

use the law of conservation of energy

$U_1 + K_1 = U_2 + K_2$

$-Gm_1m_2 / r_1 + 0 = -Gm_1m_2 / r_1 + K_2$

$-(6.67 \times 10^{-11} \text{ N-m}^2/\text{kg}^2)(2.0 \times 10^{-3} \text{ kg})^2 / (1.0 \text{ m}) =$

$-(6.67 \times 10^{-11} \text{ N-m}^2/\text{kg}^2)(2.0 \times 10^{-3} \text{ kg})^2 / (0.60 \text{ m}) + K_2$

$K_2 = 1.8 \times 10^{-16}$ J

$K = (1/2)\, mv^2$; $1.8 \times 10^{-16} = (1/2)(2.0 \times 10^{-3})\, v^2$; $v = 4.2 \times 10^{-7}$ m/s

7.6 Kepler's Laws and Earth Satelites

Example 1 Suppose there was a planet in the solar system with a period of 10 y. How far is the planet located from the Sun ?

given : $T = 10$ y ; $K = 2.97 \times 10^{-19}\ s^2 / m^3$

$T^2 = K r^3$

$(10\ y)(365\ d/y)(24\ h / d)(3600\ s / h) = 3.15 \times 10^8\ s$

$(3.15 \times 10^8)^2 = (2.97 \times 10^{-19}\ s^2 / m^3)\ r^3$; $r = 6.9 \times 10^{11}$ m

Example 2 A satellite with a mass of 1.00×10^3 kg orbits the Earth in a circular orbit whose period is 30.0 h.

A. What distance is the satellite above the Earth ?
B. What is the kinetic energy of the satellite ?
C. What is the potential energy of the satellite ?
D. What is the total energy of the satellite ?

given : $m = 1.00 \times 10^3$ kg ; $T = 30.0$ h ; $M_e = 6.0 \times 10^{24}$ kg

(a) first find the time in seconds : $30\ h\ (3600\ s/h) = 1.08 \times 10^5\ s$

the gravitational force causes the satellite to orbit : $mv^2 / r = G\ m\ M_e / r^2$

the speed of the satellite : $v = 2\pi r / t$

$4\pi^2 r^2 / r t^2 = G\ M_e / r^2$; $r = [GM_e t^2 / 4\pi^2]^{1/3}$

$r = [\ (6.67 \times 10^{-11})(6.0 \times 10^{24})(1.08 \times 10^5)^2 / 4\pi^2]^{1/3} = 4.91 \times 10^7$ m

$r = r_E + r'$; $4.91 \times 10^7 = 6.4 \times 10^6 + r'$; $r' = 4.27 \times 10^7$ m

(b) now go back and find the speed : $v = 2\pi\ r / t$; $v = 2\pi\ (4.91 \times 10^7\ m) / (1.08 \times 10^5\ s)$

$v = 2.9 \times 10^3$ m/s

$K = (1/2)mv^2$; $K = (1/2)(1.00 \times 10^3)(2.9 \times 10^3)^2 = 4.1 \times 10^9$ J

(c) potential energy with respect to the Earth and mass :

$U = -Gm_1 m_2 / r$

$U = -\ (6.67 \times 10^{-11}\ n\text{-}m^2 / kg^2)(6.0 \times 10^{24}\ kg)(1.0 \times 10^3\ kg) / (4.91 \times 10^7\ m)$

$U = -8.2 \times 10^9$ J

(d) total energy = $U + K = -8.2 \times 10^9\ J + 4.1 \times 10^9\ J = -4.1 \times 10^9$ J

Solutions to paired problems and selected problems

8. (a) $s = r\theta$; 1.0 km = (θ)(0.25 km) ; θ = 4.0 rad
(b) (4.0 rad)(360° / 2π rad) = 229°

12. $s = \theta r$; θ = 3.0 rev (2π rad/rev) = 19 rad
s = (0.45 m)(19 rad) = 8.5 m

16. a hertz (Hz is equivalent to s^{-1})

22. (a) $\omega = \theta / t$; $\omega = \pi$ rad / 150 s = 2.1×10^{-2} rad/s
(b) $v = \omega r$; v = (2.1×10^{-2} rad/s)(250 m) = 5.3 m/s

24. (a) $T_{rotation}$ = (10 h)(3600 s / h) = 3.6×10^4 s ; $f = 1 / T$
$f_{rotation}$ = 1 / (3.6×10^4 s) = 2.8×10^{-5} Hz
$T_{revolution}$ = (12 y) (365 d / y) (24 h / d) (3600 s / h) = 3.8×10^8 s
$f_{revolution}$ = 1 / (3.8×10^8 s) = 2.6×10^{-9} Hz
(b) $\omega = \theta / t$; $\omega_{rotation}$ = 2π rad / 3.6×10^4 s = 1.7×10^{-4} rad/s
$v = \omega r$; $v_{rotation}$ = (1.7×10^{-4})(6.9×10^3 km) = 1.2 km/s
$\omega_{rev} = 2\pi / T$ = 2π / (3.8×10^8 s) = 1.7×10^{-8} rad/s

29. (d) $F = mv^2 / r$; if F is increased v must increase or r must decrease

36. first find the speed of the moon as it orbits the earth
$v = 2\pi r / T$ T = (29.5 d)(24 h/d)(3600 s/h) = 2.55×10^6 s
v = [2π (3.8×10^8 m)] / (2.55×10^6 s) = 9.36×10^2 m/s
now find the acceleration : $a_c = v^2 / r$
a_c = (9.36×10^2 m/s)^2 / (3.80×10^8 m)
a_c = 2.31×10^{-3} m/s^2

39. (a) first find the speed of the mass : $v = 2\pi\, r / t$

$v = 2\,\pi\,(1.5\text{ m}) / 1.6\text{ s} = 5.9\text{ m/s}$

(b) $a_c = v^2 / r$; $a_c = (5.9\text{ m/s})^2 / (1.5\text{ m}) = 23\text{ m/s}^2$

42. (a) for the cycle not to slide down the wall, the frictional force must be equal to the weight of the cycle and driver ; the centripetal force is provided by the wall.

$f_s = \mu_s N$; $mv^2 / r = N$; $f_s = mg$; $mg = \mu_s\,(mv^2 / r)$; masses cancel

$9.8\text{ m/s}^2 = (1.1)\,v^2 / 15\text{ m}$; $v = 12\text{ m/s}$

$(12\text{ m/s})\,(3.6\text{ km/h}) = 43\text{ km/h}$; $(12\text{ m/s})\,(2.24\text{ mi/h}) = 27\text{ mi/h}$

(b) since the mass cancels, the speed is independent of the mass.

47. when the turntable stops the acceleration is zero ; if not the turntable would begin rotating in the opposite direction.

52. $\omega = \omega_o + \alpha t$; $0 = (7.65\text{ rad/s}) + \alpha\,(24.8\text{ s})$; $\alpha = -0.308\text{ rad/s}^2$

$\omega^2 = \omega_o^2 + 2\alpha\theta$; $0^2 = (7.65\text{ rad/s})^2 + 2\,(-0.308\text{ rad/s}^2)\theta$

$\theta = 95.0\text{ rad}\;(1\text{ rev} / 2\pi\text{ rad}) = 15.1\text{ rev}$

56. $\omega = \omega_o + \alpha t$; $\omega = 0 + (2.6\text{ rad/s}^2)(5.0\text{ s}) = 13\text{ rad/s}$

$(6.5\text{ rad/s})(1\text{ rev} / 2\pi\text{ rad})(60\text{ s} / \text{min}) = 1.2 \times 10^2\text{ rev/min}$

59. (a) the frictional force is not enough to make the coin move in a circle, therefore it moves off of the turntable.

(b) $\omega = (90\ \%)\,(33.3\text{ rev} / \text{min})\,(2\pi\text{ rad} / \text{rev})(1\text{ min} / 60\text{ s}) = 3.1\text{ rad/s}$

$f_s = \mu_s N$; $N = mg$; $f_s = m\omega^2 r$

$m\,\omega^2 r = \mu_s mg$; $(3.1\text{ rad/s})^2(0.10\text{ m}) = \mu_s\,(9.80\text{ m/s}^2)$; $\mu_s = 9.8 \times 10^{-2}$

65. the cup would be in accelerating reference frame ; it would experience a weightlessness effect and the water would not come out of the holes in the cup (cup and water fall at the same rate).

70. $F = G\,m_1 m_2 / r^2$

F_{12} (adjacent masses separated by a distance of 1.0 m)

$F_{12} = (6.67 \times 10^{-11}\ \text{N-m}^2/\text{kg}^2)(2.5\ \text{kg})(2.5\ \text{kg}) / (1.0\ \text{m})^2 = 4.2 \times 10^{-10}\ \text{N}$

now find the distance between masses 1 and 3 : $d^2 = (1.0\ \text{m})^2 + (1.0\ \text{m})^2$

$d = 1.4\ \text{m}$

$F_{13} = (6.67 \times 10^{-11}\ \text{N-m}^2/\text{kg}^2)(2.5\ \text{kg})(2.5\ \text{kg}) / (1.4\ \text{m})^2 = 2.1 \times 10^{-10}\ \text{N}$

now find the force in the x and in the y before finding the resultant

$F_x = (4.2 \times 10^{-10}\ \text{N}) + (2.1 \times 10^{-10}\ \text{N})(\cos 45°) = 5.7 \times 10^{-10}\ \text{N}$

$F_y = (4.2 \times 10^{-10}\ \text{N}) + (2.1 \times 10^{-10}\ \text{N})(\sin 45°) = 5.7 \times 10^{-10}\ \text{N}$

$F_r^2 = F_x^2 + F_y^2$; $F_r^2 = (5.7 \times 10^{-10}\ \text{N})^2 + (5.7 \times 10^{-10}\ \text{N})^2$

$F_r = 8.1 \times 10^{-10}\ \text{N}$ at 45°

79. $g = G\,M_m / r^2$

$g = (6.67 \times 10^{-11})(7.4 \times 10^{22}\ \text{kg}) / (1.75 \times 10^6\ \text{m})^2$

$g = 1.6\ \text{m/s}^2$

$(1.6 / 9.8) = 1 / 6$

88. $T^2 = K\,r^3$

(a) for Earth $r = 1.5 \times 10^{11}\text{m}$; $T = 365\ \text{d}\ (24\ \text{h/d})(3600\ \text{s/h}) = 3.15 \times 10^7\ \text{s}$

$(3.16 \times 10^7)^2 = K\,(1.5 \times 10^{11})^3$; $K = 3.0 \times 10^{-19}\ \text{s}^2/\text{m}^3$

(b) for Venus : $T = 1.84 \times 10^7\ \text{s}$; $r = 1.08 \times 10^{11}\ \text{m}$

$(1.84 \times 10^7)^2 = K\,(1.08 \times 10^{11})^3$; $K = 3.0 \times 10^{-19}\ \text{s}^2/\text{m}^3$

106. $v_o = 60\ \text{km/h} = 16.7\ \text{m/s}$; $v = 90\ \text{km/h} = 25.0\ \text{m/s}$

$v = v_o + at$; $25.0\ \text{m/s} = (16.7\ \text{m/s}) + a\,(10.0\ \text{s})$; $a = 0.83\ \text{m/s}^2$

$a = \alpha r$; $0.83\ \text{m/s}^2 = \alpha\,(500\ \text{m})$

$\alpha = 1.7 \times 10^{-3}\ \text{rad/s}^2$

112. (a) $x = \theta r$; $3500 \text{ km} = \theta\,(3.8 \times 10^5 \text{ km})$

$\theta = 9.21 \times 10^{-3}$ rad

(b) $12.8 \times 10^3 \text{ km} = \theta\,(3.8 \times 10^5 \text{ km})$; $\theta = 3.4 \times 10^{-2}$ rad

Sample Quiz

Multiple Choice. Choose the correct answer.

Questions 1 - 3 refer to the following.

A cylinder starts from rest and attains a speed of 4.0 rps in 2.0 s. The radius of the cylinder is 0.50 m.

__ 1. What is the angular acceleration of the cylinder ?
A. 0.20 rad/s^2 B. 2.0 rad/s^2 C. 4.0 rad/s^2 D. 12.6 rad/s^2

__ 2. How many radians does the cylinder turn during this time interval ?
A. 5.0 rad B. 12 rad C. 25 rad D. 50 rad

__ 3. What is the tangential speed of a point on the edge of the wheel at 2.0 s ?
A. 26 m/s B. 13 m/s C. 6.5 m/s D. 4.0 m/s

__ 4. A mass travel with uniform circluar motion. The radius of the circle is r and the period of the mass is 2π s. What is the speed of the mass ?
A. r m/s B. (r / π) m/s C. (πr) m/s D. ($\pi^2 r$) m/s

__ 5. A 2.0 kg mass moves in a uniform circular motion with a speed of 2.0 m/s. The radius of the circle is 0.50 m. What is the acceleration of the mass ?
A. 2.0 m/s^2 B. 4.0 m/s^2 C. 8.0 m/s^2 D. 16 m/s^2

__ 6. A hypothetical planet has a mass of half the mass of the Earth and a radius of twice the mass of the Earth. What is the acceleration due to gravity on the planet in terms of g.
A. g/16 B. g / 8 C. g / 4 D. g / 2

__ 7. The gravitational force between two masses is F. The mass of each body is doubled and the distance between the masses is halved. What is the new force present between the two masses ?
A. 2 F B. 4 F C. 8 F D. 16 F

Problems

8. A 0.50-m radius cylinder is accelerated from rest at a rate of 0.50 rad/s^2 for 5.0 s.
   A. Find the angular speed of the mass after the 5.0 s interval.
   B. Find the angular displacement during the 5.0 s interval.
   C. Find the total acceleration of the mass after the 5.0 s interval.

9. A 0.25-kg point mass is whirled in a vertical circle, radius 0.50 m. The speed at the top of the circle is just enough to keep the mass moving in the circle.
   A. Find the speed of the mass at the top of the circle.
   B. Find the speed of the mass at the bottom of the circle.

10. The mean distance from the Earth to the Sun is 1.5 x 10^{11} m and the period of the Earth's orbit around the Sun is 1.0 year. Based on this information, caculate the mass of the Sun.

Chapter 8 Rotational Motion and Equilibrium

Chapter Objectives

Upon completion of the study of rotational motion and equilibrium, students should be able to :

1. distinguish between translational and rotational motion, and write relationships for bodies which are rotating without slipping.
2. write expressions relating torque, force, and moment of inertia.
3. determine the moment of inertia when given the expression for a body, and apply the parallel axis theorem.
4. detemine the moment of inertia for a point mass.
5. write and apply expressions for the net torque acting on a rotaing body in terms of the moment of inertia and angular acceleration.
6. apply Newton's laws of motion to rotational dynamics.
7. write expressions relating work, torque, and angular displacement for uniform torques.
8. apply the law of conservation of energy to bodies which slip and roll.
9. apply the relationship between rotational work and kinetic energy.
10. distinguish between sliding friction and rolling friction.
11. write an expression which relates angular momentum to the moment of inertia and to angular velocity.
12. state and apply the law of conservation of angular momentum.
13. state the conditions for static equilibrium.
14. explain the following terms : translational equilibrium, mechanical equilibrium, rotational equilibrium, concurrent force, and static equilibrium.
15. apply the conditions for static equilibrium in choosing the pivot point properly in order to solve problems
16. describe the conditions necessary for stable equilibrium, unstable equilibrium, and neutral equilibrium.
17. solve problems applying the concept of center of mass and center of gravity.

Chapter Summary

* A **rigid body** is an object or system of particles in which the interparticle distances are fixed and remain constant.

* In pure **translational motion**, every particle of an object has the same instantaneous velocity, hence there is no rotation. In pure **rotational motion**, all the particles of a body have the same instantaneous angular velocity and travel in circles about the axis of rotation.

* The conditions for **translational** and **rotational** equilibrium are the sum of the net force and the sum of the net torque must equal zero. A body is said to be in **mechanical** equilibrium when both of these conditions are satisfied.

* For an object in **stable equilibrium** and small displacement from equilibrium results in a restoring force of torque, which tends to return the object to its original position. But for an object in **unstable equilibrium**, any small displacement results in a restoring force or torque, which tends to take the object further from its equilibrium position.

* An object is in stable equilibrium as long as its center of gravity lies vertically above and inside its original base of support.

*The condition for rolling without slipping is $v = \omega r$, where v is the speed of the center of the mass and ω is the angular speed. (Also $s = r\theta$, and for accelerated rolling $a = \alpha r$.) The axis of rotation through the point or line of contact with the surface is called the **instantaneous axis of rotation**, and the points of contact on this axis are instantaneously at rest.

*The magnitude of the **torque** produced by a force F is given by $\tau = r_{per}F$, or $\tau = rF\sin\theta$, where $r\sin\theta$ is the **moment arm** or lever **arm**, and is perpendicular distance from the axis of rotation to the line of action of the force. The units of torque are the m-N (the same as for work, N-m or J, but written in reverse for distinction.) A net torque produces a rotating acceleration.

*The torque on a particle due to a constant force is $\tau = rF_{per} = mra = mr^2\alpha$. Similarly for a rotating body with many particles $\tau = \Sigma m_i r_i^2 \alpha = I\alpha$, where $I = \Sigma m_i r_i^2$ is **moment of inertia** and represents rotational inertia (rotational analog of mass).

*The moment of inertia about a axis parallel to an axis of rotation through the center of mass is given by the **parallel axis theorem**, $I = I_{cm} + Md^2$, where d is the distance between the parallel axes and M is the total mass of the body.

*Some rotational quantities :

Work :	$W = \tau\theta$
Kinetic energy :	$K = (1/2)I\omega^2$
Power :	$P = \tau\omega$

*The kinetic energy of a **rolling body** (without slipping) relative to an axis through the center of the mass is equal to the sum of the rotational and translational kinetic energies.

***Rolling friction** comes mainly from the deformation of materials, giving rise to a mound build up in the front of the rolling object in the direction of the motion over which it must "climb".

***Angular momentum** L of a particle is given by

$$L = r_{per}p = mr_{per}v = mr^2_{per}\omega$$

and for circular motion $r_{per} = r$. For a rigid body,

$$L = (\Sigma m_i r_i^2)\omega = I\omega.$$

***Torque** may be generally expressed as the rate of change of angular momentum.

*If the net torque is zero, the change in angular momentum is zero. This is the condition for the **conservation of angular momentum**.

*For a rigid body, the moment of inertia is constant, $I = I_o$ and hence $\omega = \omega_o$. However, for a nonrigid system in which there may be a change in mass distribution, the moment of inertia may change, giving rise to a change in the angular velocity and hence an angular acceleration in the absence of a net torque.

Important Terms and Relationships

8.1 Rigid Bodies, Translations, and Rotations

rigid bodies
translational motion
rotational motion
instantaneous axis of rotation
condition of rolling without slipping $v = \omega r$; $a = \alpha r$; $s = \theta r$

8.2 Torque, Equilibrium, and Stability

moment arm or lever arm
torque $\tau = r_{per}F$ or $\tau = rF \sin\theta$
translational equilibrium $\Sigma F_i = 0$

concurrent forces

condition for rotational equilibrium $\Sigma\tau_i = 0$

mechanical equilibrium
static equilibrium
center of gravity
equilibrium
    stable
    unstable

8.3 Rotational Dynamics

rotational form of Newton's second law : $\tau = I\alpha$

moment of inertia : $I = \Sigma m_i r_i^2$

parallel axis theorem : $I = I_{cm} + Md^2$

8.4 Rotational Work and Kinetic Energy

rotational work : $W = \tau\theta$

rotational power : $P = \tau\omega$ (constant angular velocity)

rotational kinetic energy : $K = (1/2)I\omega^2$

rolling bodies : $K = (1/2)mv_{cm}^2 + (1/2)I_{cm}\omega^2$

8.5 Angular Momentum

angular momentum : $L = I\omega$

torque in terms of angular momentum : $\tau = \Delta L / \Delta t$

conservation of angular momentum : $I\omega = I_o\omega_o$ (net torques is zero)

Additional Solved Problems

8.2 Torque, Equilibrium, and Stability

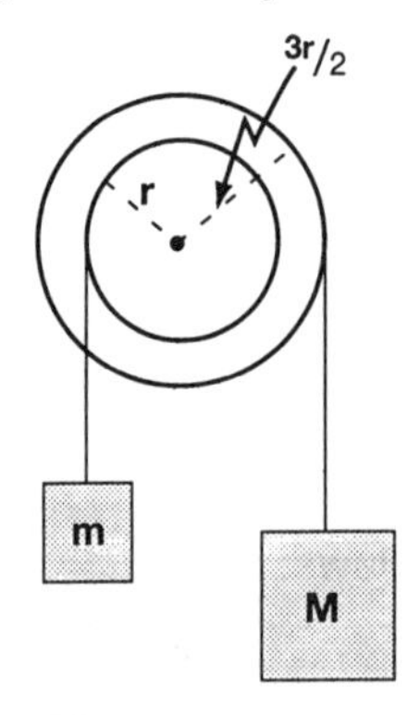

Fig. 8.1

Example 1 Find the mass m needed to keep the ideal system in equilibrium. Assume M = 6.0 kg.

the net torque should be zero, if the system is in equilibrium :

$$\tau_{cw} = \tau_{ccw}$$

$$mgr = Mg(3/2)r$$

$$m\,(9.80\ m/s^2)\,r = (6.0\ kg)(9.80\ m/s^2)(3/2)\,r$$

$$m = 9.0\ kg$$

Example 2 Find the moment of inertia for the following situations :

A. point mass of 2.0 kg located 80 cm from the reference point.

B. solid disk of mass 1.5 kg whose radius is 80 cm which is rotated about its center.

C. meter stick whose mass is 80 g which is rotated on its end.

given : (a) m = 2.0 kg ; r = 0.80 m

(b) m = 1.5 kg ; r = 0.80 m

(c) m = 0.080 kg ; r = 1.0 m

A. point mass $I = mr^2 = (2.0\ kg)(0.80\ m)^2 = 1.3\ kg\text{-}m^2$

B. solid disk about its center :

$$I = (1/2)mr^2 = (1/2)(1.5\ kg)(0.80\ m)^2 = 0.48\ kg\text{-}m^2$$

C. uniform rod pivoted about its end : $I = (1/3)mL^2$

$$I = (1/3)(0.080\ kg)(1.0\ m)^2 = 2.7 \times 10^{-2}\ kg\text{-}m^2$$

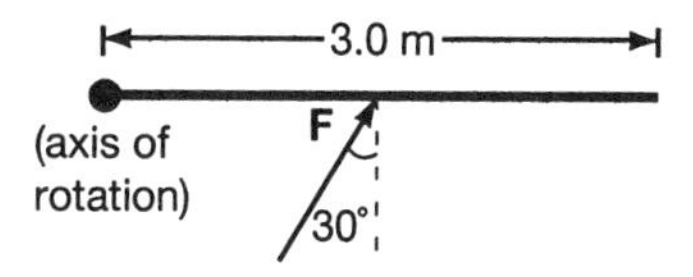

Fig. 8.2

Example 3 A force F of magnitude 10 N is applied to the midpoint of a pivoted 3.0 m bar as shown in Fig. 8.2. What is the torque exerted on the rod by the force?

$\tau = rF\ (\sin\theta) = (1.5\ m)(10\ N)(\sin 60°) = 13$ m-N

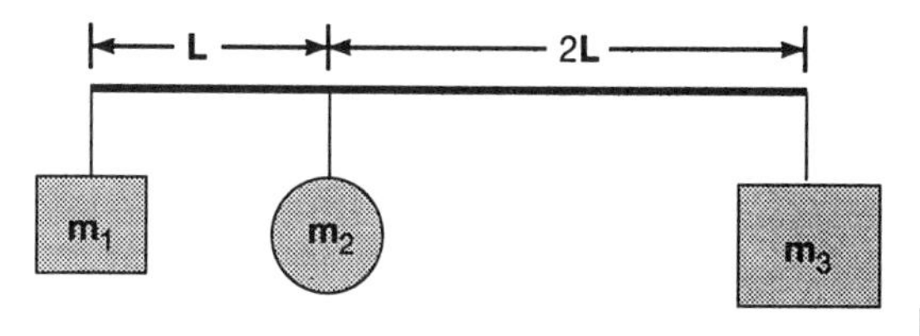

Fig. 8.3

Example 4 For the three masses suspended from a rod as shown in Fig. 8.3, find the single force (and the point of application) needed to produce translational and rotational equilibrium. Let L = 1.0 m, $m_1 = 2.0$ kg, $m_2 = 3.0$ kg, and $m_3 = 4.0$ kg.

to produce translational equilibrium the net force must equal zero--upward force needed

$F_{up} = F_{down} = m_1g + m_2g + m_3g$

$F_{up} = (2.0\ kg)(9.80\ m/s^2) + (3.0\ kg)(9.80\ m/s^2) + (4.0\ kg)(9.80\ m/s^2) = 88\ N$

to produce rotational equilibrium :

$\tau_{cw} = \tau_{ccw}$

choose the left end as a pivot :

$(3.0\ kg)(9.80\ m/s^2)(1.0\ m) + (4.0\ kg)(9.80\ m/s^2)(3.0\ m) = (88\ N)(d)$

d = 1.7 m from the left end

8.3 Rotational Dynamics

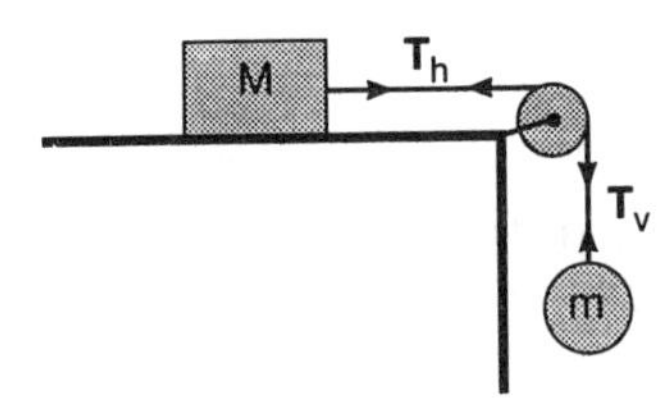

Fig. 8.4

Example 1 In the diagram above, M has a mass of 2.0 kg and m a mass of 1.0 kg. All surfaces are frictionless. The pulley has a mass of 0.20 kg and a radius of 20 cm. If the system starts from rest, neglecting pulley friction, find

A. the acceleration of the mass.
B. the angular acceleration of the pulley.
C. the tension in the horizontal string.
D. the tension in the vertical string.
E. the speed of mass M after 2.0 s.

A. using Newton's laws of motion : $Ma = T_h$

$$ma = mg - T_v$$

using torques $I\alpha = T_vR - T_hR$

$$(1/2)m_pR^2\ \alpha = = T_vR - T_hR$$

$$(1/2)m_p\ \alpha = \ mg$$

now add the three equations

$$Ma + ma + (m_p/2)a = mg$$

$$(2.0\text{ kg})\ a + (1.0\text{ kg})a + (0.10\text{ kg})\ a = (1.0\text{ kg})(9.80\text{ m/s}^2)$$

$$(3.1)a = (9.80) \quad ; \quad a = 3.2\text{ m/s}^2$$

B. $a = \alpha r$; $3.2\text{ m/s}^2 = \alpha(0.20\text{ m})$; $\alpha = 16\text{ rad/s}^2$

C. $Ma = T_h$; $(2.0\text{ kg})(3.2\text{ m/s}^2) = T_h = 6.4\text{ N}$

D. $ma = mg - T_v$; $(1.0\text{ kg})(3.2\text{ m/s}^2) = (1.0\text{ kg})(9.80\text{ m/s}^2) - T_v$; $T_v = 6.6\text{ N}$

E. $v = v_o + at$; $v = 0 + (3.2\text{ m/s}^2)(2.0\text{ s}) = 6.4\text{ m/s}$

Example 2 The switch on a fan rotating at 100 rpm is shut off and the fan is observed to comeuniformly to a halt in 20.0 s. If the fan has a moment of inertia of 0.80 kg-m^2, find

A. the angular acceleration of the fan.
B. the frictional torques stops the fan.

given : ω_o =100 rpm = 10.5 rad/s ; $\omega = 0$; t = 20.0 s ; I = 0.80 kg-m^2

A. first find the angular acceleration

$\omega = \omega_o + \alpha t$; 0 = 10.5 rad /s + α (20.0 s) ; α = -0.53 rad/s^2

B. $\tau = I\alpha$; τ = (0.80 kg-m^2)(-0.53 rad/s^2) = - 0.42 m-N

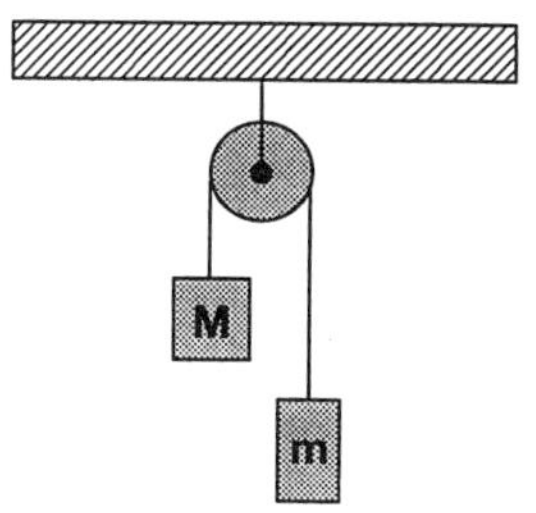

Fig. 8.5

Example 3 In Fig. 8.5, the mass M is 200 g and m is 100 g. The pulley has a mass of 150 g and a radius of 20 cm. I for the pulley = (1/2)mR^2

A. Find the linear acceleration of the system.

B. Find the angular acceleration of the pulley.

(a) first use Newton's law of motion : $Ma = Mg - T_L$; $ma = T_r - mg$

$I\alpha = Mgr - mgr$; $(1/2)mr^2\,\alpha = Mgr - mgr$

now add the three equations :

$Ma + ma + (M_p/2)a = Mg - mg$

(0.200 kg)a + (0.100 kg) a + (0.150 kg / 2) a = (0.200 kg)(9.80) - (0.100)(9.8)

a = 2.6 m/s^2

(b) $a = \alpha r$; 2.6 m/s^2 = α (0.20 m) ; α = 13 rad/s^2

8.4 Rotational Work and Kinetic Energy

Example 1 A uniform solid disk is free to rotate about an axis through its center. Initially the disk is at rest and its angular speed is increased to 100 rpm. If the moment of inertia is 0.50 kg-m^2, what is

A. the work done to bring the cylinder to its new speed ?

B. the change in kinetic energy.

given : I = 0.50 kg-m^2 ; $\omega_o = 0$; ω = 100 rpm = 10.5 rad/s

for A and B $W = \Delta K$ = (1/2)(0.50 kg-m^2)(10.5 rad/s)^2 = 28 J

Example 2 A yo-yo [solid disk -- mass 0.100 kg, radius 0.10 m, $I = (1/2)mR^2$], is released from rest. As it descends the yo-yo unwinds and does not slip. When the mass descends 1.0 m, find

A. the linear speed of the yo-yo.
B. the angular speed of the yo-yo.
C. the linear acceleration of the yo-yo as it moves the 1.0 m.

A. $U_1 + K_1 = U_2 + K_2$ (1 -- at top ; 2 -- 1.0 m lower)

$mgh + 0 = 0 + (1/2)mv^2 + (1/2)\,I\omega^2$

$mgh = (1/2)\,mv^2 + (1/2)(1/2\,mR^2)\,\omega^2$

$mgh = 1/2\,mv^2 + (1/4)\,mv^2$

$(9.80\text{ m/s}^2)(1.0\text{ m}) = (3/4)\,v^2$; $v = 3.6\text{ m/s}$

B. $v = r\omega$; $3.6\text{ m/s} = (0.10\text{ m})\,\omega$; $\omega = 36\text{ rad/s}$

C. $v^2 = v_o^2 + 2ax$

$(3.6\text{ m/s})^2 = 0^2 + 2\,a\,(1.0\text{ m})$; $a = 6.5\text{ m/s}^2$

Example 3 A solid cylinder [$I = (1/2)mR^2$] rolls down a 30° inclined plane without slipping. The plane is 10 m long and the radius of the cylinder is 10 cm.

A. Find the linear speed of the cylinder at the bottom of the plane.
B. Find the angular speed of the cylinder at the bottom of the circle.
C. Find the acceleration of the mass as it moves down the plane.

A. $U_1 + K_1 = U_2 + K_2$

$mgh + 0 = (1/2)\,mv^2 + (1/2)I\omega^2$

$mgh = (3/4)mv^2$

$m\,(9.80\text{ m/s}^2)(5.0\text{ m}) = (3/4)\,m\,v^2$; $v = 8.1\text{ m/s}$

B. $v = r\omega$; $8.1\text{ m/s} = (0.10\text{ m})\omega$; $\omega = 81\text{ rad/s}$

C. $v^2 = v_o^2 + 2ax$; $(8.1\text{ m/s})^2 = 0^2 + 2(a)(10\text{ m})$; $a = 3.3\text{ m/s}^2$

8.5 Angular Momentum

Example 1 Find the angular momentum of the moon as it moves around the Earth. The mass of the moon is 7.4×10^{22} kg and the mean distance from the Earth to the moon is 3.8×10^{8} m and the time for the moon to orbit the Earth is 29.5 days.

$L = I\omega$; $I = mr^2 = (7.4 \times 10^{22}\text{ kg})(3.8 \times 10^8)^2 = 1.1 \times 10^{40}\text{ kg-m}^2$

$L = (1.1 \times 10^{40}\text{ kg-m}^2)\,[2\pi\,(3.8 \times 10^8)]\,/\,(2.55 \times 10^6\text{ s}) = 1.0 \times 10^{43}\text{ N-m/s}$

Example 2 A diver springs from a board and begins rotating with an angular speed of 1.5 rps. At this point his moment of inertia is 90 kg-m^2. He then decreases his moment of inertia to 60 kg-m^2.
A. What is his new angular speed ?
B. What is the change in kinetic energy ?

given : $I_o = 90\text{ kg-m}^2$; $I = 60\text{ kg-m}^2$; $\omega_o = 1.5\text{ rps} = 9.42\text{ rad/s}$

A. angular momentum is conserved :

$L = L_o$

$I\omega = I_o\omega_o$

$(90\text{ kg-m}^2)(9.42\text{ rad/s}) = (60\text{ kg-m}^2)\,\omega$; $\omega = 14\text{ rad/s}$

B. $K_o = (1/2)\,I_o\omega_o^2 = (1/2)\,(90)(9.42)^2 = 4.0 \times 10^3\text{ J}$

$K = (1/2)(60)(14)^2 = 5.9 \times 10^3\text{ J}$; $\Delta K = 1.9 \times 10^3\text{ J}$

Solutions to paired problems and selcted problems

4. If $v < R\omega$, the tires are partially spinning without corresponding forward motion. If $v > R\omega$, the cylinder is not in constant contact with the ground -- it is "skipping forward".

8. $v = r\omega$; $v = (2.4\text{ rad/s})(0.20\text{ m}) = 0.48\text{ m/s}$; $x = vt$; $x = (0.48\text{ m/s})(2.0\text{ s}) = 0.96\text{ m}$ the center of the cylinder moves 96 cm, therefore since the distances are equal the cylinder does not slip.

13. (a) no , since the forces must be perpendicular (or a component perpedicular to the axis of rotation
(b) yes, because translational equilibrium can exist without having rotational equilibrium
(c) no
(d) no

22. (a) the net torque should be zero : $m_1gr_1 = m_2gr_2$

$(100\text{ g})(9.80\text{ m/s}^2)(25.0\text{ cm}) = (75.0\text{ g})(9.80\text{ m/s}^2)(x)$
x = 33.3 cm to the right of the 50 cm mark or at the 83.3 cm mark

(b) $m_1gr_1 = m_2gr_2$

$(100\text{ g})(9.80\text{ m/s}^2)(25.0\text{ cm}) = m_2\,(9.80\text{ m/s}^2)(40.0\text{ cm})$; $m_2 = 62.5\text{ g}$

26. the net torque should be zero : $F_1r_1 = F_2r_2$

$F_m\,(10\text{ cm})(\sin 15°) = (3.0\text{ kg})(9.80\text{m/s}^2)(26\text{ cm})$; $F_m = 3.0 \times 10^2$ N

32. $T_v = (1.5\text{ kg})(9.80\text{ m/s}^2) = 15$ N

$T_{up}(\cos 45°) = T_2\,(\cos 30°)$; $T_{up}\,(\sin 45°) = T_2 \sin 30° + 15$ N

$T_2\,(\cos 30° - \sin 30°) = 15$ N

$T_2\,(0.37) = 15$

$T_2 = 41$ N ; $T_1 = 50$ N

38. $w_px = w_sx_{CM}$; x = (15 kg / 70 kg)(1.25 m) = 0.27 m ; hence 0.27 m from the left of the left support is unsafe.

46. (a) the meterstick has a higher center of mass ; a larger moment of inertia and smaller angular acceleration, making it easier to balance.
(b) softball wins because of mass distribution ; others tie

54. first find the moment of inertia :

$I = (1/2)mr^2 = (1/2)(2.0 \times 10^3\text{ kg})(30\text{ m})^2 = 9.0 \times 10^5\text{ kg-m}^2$

now find the angular accleration

$\omega = \omega_0 + \alpha t$; $1.5\text{ rad/s} = 0 + \alpha\,(12\text{ s})$; $\alpha = 1.3 \times 10^{-1}\text{ rad/s}^2$

finally the net torque can be found $\tau = I\alpha = (9.0 \times 10^5\text{ kg-m}^2)(0.13\text{ rad/s}^2)$

$\tau = 1.2 \times 10^5$ m-N

58. first find the moment of inertia: $I = (1/2)mr^2 = (1/2)(2.0 \text{ kg})(0.75 \text{ m})^2 = 0.56 \text{ kg-m}^2$
$\tau = I\alpha = (0.56 \text{ kg-m}^2)(4.8 \text{ rad/s}^2) = 2.7 \text{ m-N}$; $\tau = Fr$; $2.7 \text{ m-N} = F\,(0.75 \text{ m})$;
$F = 3.6 \text{ N}$

68. use the law of conservation of energy $U_1 + K_1 = U_2 + K_2$

$0 + (1/2)\, mv^2 + (1/2)\, I\omega^2 = mgh$; $I = (2/5)mr^2$
$(1/2)\, mv^2 + (1/2)(2/5)\, mr^2\, (v/r)^2 = mgh$ (since $v = \omega r$)
$(7/10)\, mv^2 = mgh$ $v = (8.0 \text{ rad/s})(0.15 \text{ m}) = 1.2 \text{ m/s}$
$(7/10)(1.2)^2 = (9.80 \text{ m/s}^2)\, h$
$h = 0.10 \text{ m}$

76. first the moment of inertia : $I = (1/3)\, mL^2$
using the law of conservation of energy :
$U_1 + K_1 = U_2 + K_2$
$mg\,(L/2) + 0 = 0 + (1/2)\, I\omega^2$; $mg\,(L/2) = (1/2)(1/3)mL^2\, \omega^2$
$(9.80 \text{ m/s}^2)(0.10 \text{ m}) = (1/6)(0.20 \text{ m})^2\, \omega^2$; $\omega = 12 \text{ rad/s}$
$v = \omega r$; $v = (12 \text{ rad/s})(0.20 \text{ m}) = 2.4 \text{ m/s}$

79. (c) angular momentum does not change with time

88. $L = I\omega =$; $I = mr^2 = (6.0 \times 10^{24} \text{ kg})(1.5 \times 10^{11} \text{ m})^2 = 1.35 \times 10^{47} \text{ kg-m}^2$
$L_{orbital} = (1.35 \times 10^{47} \text{ kg-m}^2)(\, 2\pi \text{ rad} / 3.16 \times 10^7 \text{ s}) = 2.7 \times 10^{40} \text{ m-N-s}$

now for spinning :
$I = (2/5)mr^2 = (2/5)(6.0 \times 10^{24} \text{ kg})(6.4 \times 10^6 \text{ m})^2 = 9.83 \times 10^{37} \text{ kg-m}^2$
$L_{spin} = (9.38 \times 10^{37} \text{ kg-m}^2)\,(\, 2\pi \text{ rad} / 8.64 \times 10^4 \text{ s}) = 7.1 \times 10^{33} \text{ m-N-s}$
the vectors are not in the same direction

92. angular momentum is conserved :
$L_o = L$
$I_o\omega_o = I\omega$; $(100 \text{ kg-m}^2)(2.0 \text{ rps}) = (75 \text{ kg-m}^2)\, \omega$; $\omega = 2.7 \text{ rad/s}$

98. $I = \Sigma\, m_i r_i^{\,2} = \Sigma_1 m_1 r_1^{\,2} + \Sigma\, m_2 r_2^{\,2} = I_1 + I_2$

103. (a) $v = \omega r$; $1.5 \text{ m/s} = \omega (0.40 \text{ m})$; $\omega = 3.8 \text{ rad/s}$

(b) $I = mr^2$; $I = (2.0 \text{ kg})(0.40 \text{ m})^2 = 0.32 \text{ kg-m}^2$

$L = I\omega = (0.32 \text{ kg-m}^2)(3.8 \text{ rad/s}) = 1.2 \text{ m-N-s}$

Sample Quiz

Multiple Choice. Choose the correct answer.

__ 1. The edge of a rotating sphere moves with a linear speed of 2.0 m/s. The radius of the sphere is 0.50 m. What is the angular speed of the sphere if the sphere does not slip ?
A. 0.50 rad/s B. 1.0 rad/s C. 4.0 rad/s D. 8.0 rad/s

__ 2. The rotational analog of mass is
A. force B. torque C. inertia D. kinetic energy

__ 3. Which of the following is not a conservative law in mechanics ?
A. law of conservation of linear momentum
B. law of conservation of energy
C. law of conservation of angular momentum
D. law of conservation of forces

__ 4. A 1.2 N-force is applied 0.50 m from the center of a disk. If the force is applied perpendicular to the edge of the disk, the net torque will be
A. 0.042 m-N B. 0.60 m-N C. 24 m-N D. 72 m-N

__ 5. A solid disk and a hollow disk, have indentical masses and radii. The masses have the same intial speed. If they roll without slipping, which mass will travel the greater distance ?
A. solid disk B. hollow disk C. the distance is the same

__ 6. Two identical masses are releases from rest on an inclined plane. Mass m_1 rolls down the plane and mass m_2 slides down the inclined plane. Which mass will arrive at the bottom of the plane first ?
A. m_1 B. m_2 C. they arrive at the same time

__ 7. A skater is rotating with an angular velocity when she holds her hands inward. Which of the following is correct concerning the kinetic energy and the angular momentum ?

	Kinetic energy	angular momentum
A.	increases	increases
B.	increases	remains the same
C.	remains the same	increases
D.	remains the same	remains the same

Problems. Work the following.

8. A uniform meterstick (mass 100 g) is suspended from both ends by spring scales. A 200-g mass is placed at the 20 cm mark and a 75-g mass is placed at the 90 cm mark. What is the reading of each scale ?

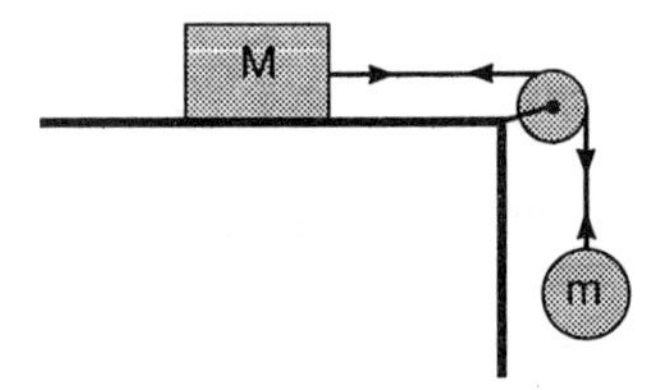

Fig. 8.7

9. In the system shown in Fig. 8.7, the mass is released from rest. All surfaces are frictionless. The radius of the pulley is 0.20 m and the mass of the pulley is 0.50 kg. The mass M is 1.0 kg and the mass m is 0.80 kg. $I = (1/2)mr^2$ for the pulley
A. Find the angular acceleration of the pulley.
B. Find the angular momentum of the pulley after 5.0 m.
C. Find the decrease in graviational potential energy after 5.0 s.
D. Find the kinetic energy of the pulley, m, and M after 5.0 s.

10. A 2.0-kg (radius 10-cm) sphere rolls down a 37° inclined plane 10 m long without slipping. The mass is released from rest at the top of the plane.
A. Find the linear speed of the mass at the bottom of the plane.
B. Find the angular speed of the mass at the bottom of the plane.
C. Find the linear acceleration of the mass at the bottom of the plane.
D. Describe what would occur if the object slides down the plane.

CHAPTER 9 Solids and Fluids

Chapter Objectives

Upon completion of the unit on solids and fluids, students should be able to :

1. define stress, strain, and elastic modulus.
2. apply the concepts of stress, strain, and modulus to problem solving.
3. interpret a graph of the stress versus the strain in order to find the slope, and recognize the proportional limit, the elastic limit, and the parts which obey Hooke's law.
4. state and apply to problem solving the appropriate type of elastic moduli.
5. find the compressibility using the bulk modulus.
6. compare and contrast the properties of solids, liquids, and gases in terms of their shape, volume, density, and compressibility.
7. state and apply the definition of a fluid in terms of shear stress.
8. define the pressure in terms of force and area, and explain the effects in the force and area on the pressure.
9. list units used to measure pressure and convert units of pressure.
10. write an expression for the variation of pressure associated with depth and apply the expression to solve problems.
11. solve problems on fluids at rest.
12. state and apply Pascal's principle.
13. explain the concept of buoyancy, and calculate the buoyant force on an object submerged in a liquid.
14. explain and apply Archimedes' principle.
15. compare the density of an object to the density of a fluid to determine if the object will sink or float , and how much of the material will be above or below the surface.
16. determine the specific gravity of a substance.
17. describe the effects of surface tension and capillary action, and explain how they are affected by differing cohesive forces.
18. calculate the contact angle and the change in height due to interactive forces.
19. state the four characteristics of an ideal fluid.
20. state and apply the flow rate equation.
21. relate the law of conservation of energy to Bernoulli's equation in order to solve problems of fluid dynamics.
22. explain and apply properies for fluids which are not ideal, including viscosity, Poiseuille's law, and the Reynold's number.

Chapter Summary

*All materials are elastic to some degree and can be deformed. **Stress** is the quantity that describes the force causing a deformation. **Strain** is a relative measure of how much deformation a given stress produces. Quantitatively, stress is the applied force per unit cross-sectional area, and strain is the ratio of the change in dimensions to the original dimensions.

*In general, stress is proportional to the strain. The constant of proportionality, which depends on the nature of the material, is called the **elastic modulus** and is the ratio of the stress to strain.

*Three types types of moduli are respectively : Young's modulus, the shear modulus, and the bulk modulus. The compressibility is the inverse of the bulk modulus.

*A **fluid** is a substance that can flow and cannot support a shear. Both liquids and gases are fluids.

***Pressure** is the force per unit area ($p = F/A$) and has units of N/m^2 or Pa.

*The pressure on an object submerged in a liquid is given by the **pressure-depth equation**, $p = p_o + \rho gh$, where p_o is the pressure on the liquid surface, ρ is the density, and h is the depth below the surface.

***Pascal's principle** states that the pressure applied to an enclosed fluid is transmitted undeminished to every point in the fluid and to the walls of its container.

*The pressure measured by an open-tube manometer is $p = p_a + \rho gh$, where p_a is the atmospheric pressure, ρ is the density of the fluid, and h is the difference in the heights of the liquid levels in the manometer arms. The pressure p is the absolute pressure; $p - p_a$ is called the gauge pressure. A closed-tube manometer measures gauge pressure. A barometer is a closed-tube manometer used to measure atmospheric pressure.

***Archimedes' principle** states that a body immersed wholly or partially in a fluid is buoyed up by a force equal in magnitude to the weight of the volume it displaces.

An object will float in a fluid if the density of the object is less than the density of the fluid.
An object will sink in a fluid if the density of the object is greater than the density of the fluid.
An object will be in equilibrium at any submerged depth in a fluid if densities of the object and the fluid are equal.

*The **specific gravity** of a substance is the ratio of the weight of a volume of the substance to the weight of an equal volume of water. This is equal to the ratio of the densities of the substance and water.

*The **surface tension** of a liquid is caused by the inward pull on the molecules of the surface layer which causes the surface to contract.

***Adhesive forces** (or adhesion) are attractive forces between unlike molecules. **Cohesive forces** (or cohesion) are attractive forces between like molecules. A relative measure of these is given by the contact angle which is the angle between a surface and a line tangent to a drop of liquid on a surface.

*The rise of a liquid in a small diameter tube is due to **capillary action**, a result of both surface tension and adhesion.

*Ideal fluid flow is steady, irrotational, nonviscous, and incompressible. For the conservation of mass, the **equation of continuity** for fluid flow in a uniform tube is ρAv = constant, where ρ is the fluid density, A is the cross-sectional area of the tube, and v is the flow speed. For an incompressible fluid, Av = constant, which is called the **flow rate equation**.

*Using the law of conservation of work and energy a relationship for an ideal fluid flow is known as **Bernoulli's** equation : $p + (1/2)\rho v^2 + \rho gy = \text{constant}$.

*Viscosity is a measure of the fluid's internal resistance to flow.

*The average **flow rate**, $Q = Av$, depends on characteristics of the fluid and the pipe as well as on the pressure difference between the ends of the pipe. **Poiseulle's law** gives a relationship for the flow rate.

*The **Reynolds number** is a dimensionless number whose value gives an indication of when turbulence will occur in fluid flow.

Important Terms and Relationships

Introducation

fluid

9.1 Solids and Elastic Moduli

stress : stress = F / A
strain : strain = $\Delta L / L_o$
tensile stress
tensile strain
elastic moduli
elastic limit
Young's modulus : $Y = (F / A) / (\Delta L / L_o)$
shear stress
shear strain
shear angle
shear modulus : $S = (F / A) / (x/h) = (F/A) / \phi$
volume stress
volume strain
bulk modulus : $B = (F/A) / -(\Delta V/V_o) = -\Delta p / (\Delta V/V_o)$
compressibility : $k = 1 / B$

9.2 Fluids : Pressure and Pascal's Principle

pressure : $p = F / A$
pascal (Pa)
atmosphere (atm)
pressure - depth equation : $p = \rho g h$
total pressure : $p = p_o + \rho g h$
Pascal's principle
open-tube manometer
absolute pressure
gauge pressure
closed-tube manometer
barometer
standard atmosphere
torr

9.3 Buoyancy and Archimedes' Principle

buoyant force : $F_b = \rho_f g V_f$
Archimedes' principle
specific gravity : sp.gr. $= \rho_s / \rho_{water}$
hydrometer

9.4 Surface Tension and Capillary Action

surface tension : $\gamma = F / L$
adhesive forces
cohesive forces
contact angle
capillary height : $h = 2\gamma(\cos\phi) / \rho g r$
capillary action

9.5 Fluid Dynamics and Bernoulli's Equation

steady flow
streamlines
irrotational flow
viscous flow
incompressible flow
equations of continuity : $\rho_1 A_1 v_1 = \rho_2 A_2 v_2$
flow rate (for an incompressible fluid) : $A_1 v_1 = A_2 v_2$
Bernoulli's principle : $p_1 + \rho g y_1 + (1/2)\rho v_1^2 = p_2 + \rho g y_2 + (1/2)\rho v_2^2$

9.6 Viscosity, Poiseuille's Law, and Reynolds Number

viscosity
coefficient of viscosity : $\eta = Fh / Av$
flow rate $Q = Av = V / t$
poiseuille (Pl)
poise (P)
Poiseuille's law : $Q = \pi r^4 \Delta p / 8\eta L$
Reynolds number : $R_n = \rho v d / \eta$

Additional Solved Problems

9.1 Solids and Elastic Moduli

Example 1 A steel wire 2.0 m in length and 2.0 mm in diameter supports a 10.0 kg mass.
A. What is the stress in the wire ?
B. What is the distance the wire is stretched ?

given : $Y_{steel} = 20 \times 10^{10}$ N/m^2 ; L = 2.0 m ; d = 2.0×10^{-3} m ; m = 10.0 kg

A. stress = F / A ; $A = \pi d^2 / 4$; A = $(3.14)(2.0 \times 10^{-3}\ \text{m})^2 / 4 = 3.1 \times 10^{-6}\ \text{m}^2$
stress = $(10.0\ \text{kg})(9.80\ \text{m/s}^2) / (3.1 \times 10^{-6}\ \text{m}^2) = 3.2 \times 10^7\ \text{N/m}^2$

B. $Y = (F/A) / (\Delta L/L_o)$; $(20 \times 10^{10}\ \text{N/m}^2) = (3.2 \times 10^7\ \text{N/m}^2) / (\Delta L/L_o)$
$\Delta L / L_o = 1.6 \times 10^{-4}$; $\Delta L = (1.6 \times 10^{-4})(2.0\ \text{m}) = 3.2 \times 10^{-4}$ m or 0.32 mm

Example 2 A shear force of 4.0×10^2 N is applied to one face of a steel cube with sides measuring 20 cm. What is the relative displacement ?

given : F = 4.0×10^2 N ; $S_{steel} = 8.2 \times 10^{10}$ N/m^2 ; h = 20 cm
$S = (F / A) / (x / h)$
$F / A = (4.0 \times 10^2\ \text{N}) / (0.20\ \text{m})^2 = 1.0 \times 10^4\ \text{N/m}^2$
$8.2 \times 10^{10}\ \text{N/m}^2 = (1.0 \times 10^4\ \text{N/m}^2) / (x / h)$
$(x / h) = 1.2 \times 10^{-7}$
$x / (20\ \text{cm}) = (1.2 \times 10^{-7})$; $x = 2.4 \times 10^{-6}$ cm

9.2 Fluids : Pressure and Pascal's Principle

Example 1 A pool is filled with water 3.0 m deep.
A. What is the total pressure at this depth ?
B. What is the force on a rectangular drain (20 cm by 30 cm) ?

given : $\rho = 1.0 \times 10^3$ kg/m^3 ; h = 3.0 m ; A = 0.20 m x 0.30 m
A. $p = p_a + \rho gh$; $p = (1.013 \times 10^5\ \text{Pa}) + (1.0 \times 10^3\ \text{kg/m}^3)(9.80\ \text{m/s}^2)(3.0\ \text{m})$
$p = 1.31 \times 10^5\ \text{N/m}^2$

B. $p = F / A$; $A = (0.20 \text{ m})(0.30 \text{ m}) = 6.0 \times 10^{-2} \text{ m}^2$
$(1.31 \times 10^5 \text{ N/m}^2) = F / (6.0 \times 10^{-2} \text{ m}^2)$; $F = 7.9 \times 10^3 \text{ N}$

Example 2 A 0.20-kg steel cube is submerged at the bottom of a large reservior of water whose depth is 10 m. What is the force on the cube due to the pressure of the water and air ?

given : $\rho_{water} = 1.0 \times 10^3 \text{ kg/m}^3$; $m = 0.20 \text{ kg}$; $\rho_{steel} = 7.8 \times 10^3 \text{ kg/m}^3$

first find the pressure at this location : $p = p_a + \rho gh$

$p = (1.013 \times 10^5 \text{ N/m}^2) + (1.0 \times 10^3 \text{ kg/m}^3)(9.80 \text{ m/s}^2)(10 \text{ m}) = 1.99 \times 10^5 \text{ N/m}^2$

$p = F / A$; the area must be found.

$\rho = m / V$; $(7.8 \times 10^3 \text{ N/m}^2) = (0.20 \text{ kg}) / V$; $V = 2.6 \times 10^{-5} \text{ m}^3$

$V = s^3$; $2.6 \times 10^{-5} \text{ m}^3 = s^3$; $s = 2.9 \times 10^{-2} \text{ m}$

$A = s^2 = (2.9 \times 10^{-2} \text{ m})^2 = 8.4 \times 10^{-4} \text{ m}^2$

$(1.99 \times 10^5 \text{ N/m}^2) = F / (8.4 \times 10^{-4} \text{ m}^2)$; $F = 1.7 \times 10^2 \text{ N}$

9.3 Buoyancy and Archimedes' Principle

Example 1 A rectangular solid has a mass of 100 kg and a volume of 0.25 m^3. The solid is placed into water. Will it sink or float, and if it sinks what will be its apparent weight ?

given : $m = 100 \text{ kg}$; $V = 0.25 \text{ m}^3$

$\rho = m / V$; $\rho = (100 \text{ kg}) / (0.25 \text{ m}^3) = 400 \text{ kg/m}^3$;
since the density of the solid is less than water, the mass will float and its apparent weight will be zero.

Example 2 A downward force of 50 N is needed to hold a 0.010-m^3 block submerged in water and a 20 N force is required in an unknown liquid. What is the density of the block and the liquid ?

given : $F_{down\ (1)} = 50 \text{ N}$; $V = 0.010 \text{ m}^3$; $\rho_{water} = 1.0 \times 10^3 \text{ kg/m}^3$;
$F_{down\ (2)} = 20 \text{ N}$

$F_b = \rho_f g V_f = (1.0 \times 10^3 \text{ kg/m}^3)(9.80 \text{ m/s}^2)(0.010 \text{ m}^3) = 98 \text{ N}$
$F_{up} = F_{down}$; $F_b = 50 \text{ N} + mg$; $98 \text{ N} = 50 \text{ N} + mg$; $mg = 48 \text{ N}$
$mg = 48 \text{ N}$; $m = 4.9 \text{ kg}$
$\rho = m / V$; $\rho = (4.9 \text{ kg}) / (0.010 \text{ m}^3) = 4.9 \times 10^2 \text{ kg/m}^3$

$F_b = F_{down} + mg = 20 \text{ N} + 48 \text{ N} = 68 \text{ N}$
$68 \text{ N} = \rho_f V_f g$; $68 \text{ N} = \rho_f (0.010 \text{ m}^3)(9.80 \text{ m/s}^2)$; $\rho_f = 6.9 \times 10^2 \text{ kg/m}^3$

9.4 Surface Tension and Capillary Action

Example To what height would whole blood rise in a glass capillary tube whose
(a) diameter is 0.50 mm ?
(b) diameter is 1.00 mm ?

given : $Y = 0.058 \text{ N/m}$; $d_1 = 0.50 \text{ mm}$; $d_2 = 1.00 \text{ mm}$

(a) $h_b = 2\gamma / \rho g r$

$h_b = 2 (0.058 \text{ N/m}) / [(1.05 \times 10^3 \text{ kg / m}^3)(2.5 \times 10^{-4} \text{ m}) = 0.44 \text{ m}$

(b) $h_b = 2 (0.058 \text{ N/m}) / [(1.05 \times 10^3 \text{ kg/m}^3)(5.0 \times 10^{-4} \text{ m}) = 0.22 \text{ m}$

9.5 Fluid Dynamics and Bernoulli's Equation

Example 1 Air moves across a surface area of 20 m^2 with a speed of 30 m/s. What force is produced on the area due to the change in pressure ?

given : $\rho_{air} = 1.29 \text{ kg/m}^3$; $A = 20 \text{ m}^2$; $v = 30 \text{ m/s}$

$\Delta p = (1/2) \rho v^2 = (1/2)(1.29 \text{ kg/m}^3)(30 \text{ m/s})^2 = 5.8 \times 10^2 \text{ Pa}$
$p = F / A$
$5.8 \times 10^2 \text{ N/m}^2 = F / (20 \text{ m}^2)$; $F = 1.2 \times 10^4 \text{ N}$

Example 2 Water flows from a 2.0 cm diameter horizontal pipe with a speed of 0.50 m/s into a 1.0-cm diameter pipe.
A. What is the speed of the water in the smaller pipe ?
B. What is the difference in pressure between the pipes ?

given : $d_1 = 2.0$ cm ; $v_1 = 0.50$ m/s ; $d_2 = 1.0$ cm ; ρ_{water} 1.0×10^3 kg/m^3

A. $A_1 v_1 = A_2 v_2$

$(\pi d_1{}^2 / 4)\, v_1 = (\pi d_2{}^2 / 4)\, v_2$

$(d_1 / d_2)^2 = v_2 / v_1$

$(2.0 / 1.0)^2 = v_2 / (0.50 \text{ m/s})$; $v_2 = 2.0$ m/s

B. $p_1 + (1/2)\, \rho v_1{}^2 = p_2 + (1/2)\, \rho\, v_2{}^2$

$p_1 + (1/2)(1.29 \text{ kg/m}^3)(0.50 \text{ m/s})^2 = p_2 + (1/2)(1.29 \text{ kg/m}^3)(2.0 \text{ m/s})^2$

$p_1 - p_2 = 2.42$ Pa

9.6 Viscosity, Poiseuille's Law, and Reynolds Number

Example The diameter of a blood vessel decreases to half of its original value. By what factor would the pressure change ?

$Q = \pi r^4 \Delta p \;/ 8\, \eta\, L$; $r_1{}^4\, \Delta p_1 = r_2{}^4\, \Delta p_2$; $r^4 \Delta p_1 = [(1/2) r]^4\, \Delta p_2$; $p_2 = 16\, p_1$

Solutions to paired problems and selected problems

2. (a) shear moduli exist only for solids

10. $Y = (F / A) / (\Delta L / L_o)$;

$Y = \{ [(6.0 \text{ kg})(9.80 \text{ m/s}^2)\,] / [\,(\pi)(0.50 \times 10^{-3} \text{ m})^2\,]\,\} / [\,(1.4 \times 10^{-3} \text{ m}) / (2.0 \text{ m})]$

$Y = 1.1 \times 10^{11}$ N/m^2

20. $p = -\,\rho \Delta V / \beta$; $V = V_o\,(1 - \rho / \beta)$

$V = (6.0 \times 10^{-2} \text{ m})^3 \{\, 1 - [(1.2 \times 10^{-7} \text{ kg/m}^3) / (7.5 \times 10^{10} \text{ N/m}^2)]\}$

$V = 2.15965 \times 10^{-4}$ m^3

26. the difference in lengths does not effect the pressure on the dam ; since the depth is the same ; the pressure is the same.

32. $\rho = m / V$

Al : $(2.7 \times 10^3 \text{ kg/m}^3) = (10 \text{ kg}) / V$; $V = 3.7 \times 10^{-3}$ m^3

Pb : $(11.3 \times 10^3 \text{ kg / m}^3) = (2.4 \text{ kg}) / V$; $V = 2.1 \times 10^{-4}$ m^3 ; $V_{Al} = (18)\, V_{Pb}$

36. $p = \rho gh$; $p = (1.0 \times 10^3 \text{ kg/m}^3)(9.80 \text{ m/s}^2)(12 \text{ m}) = 1.2 \times 10^5 \text{ Pa}$
$A = \pi d^2 / 4$; $A = \pi\ (0.18 \text{ m})^2 / 4 = 2.5 \times 10^{-2} \text{ m}^2$
$P = F / A$; $(1.2 \times 10^5 \text{ Pa}) = F / (2.5 \times 10^{-2} \text{ m}^2)$; $F = 3.0 \times 10^3 \text{ N}$

42. $\Delta p = \rho gh$; $\Delta p = (13.6 \times 10^3 \text{ kg/m}^3)(9.80 \text{ m/s}^2)(0.050 \text{ m}) = 6.7 \times 10^3 \text{ Pa}$
$\Delta p = \rho g h_{water}$; $6.7 \times 10^3 \text{ Pa} = (1.0 \times 10^3 \text{ kg/m}^3)(9.80 \text{ m/s}^2)(h_{water})$
$h_{water} = 0.68 \text{ m}$ or 68 cm

51. There is no change. By Archimedes' principle, the weight of the floating ice cube equals the weight of the water displaced. Hence the ice has the same mass or quantity of matter as the displaced water. Assuming the mass of air is negligible, same for hollow cubes, but less water is displaced by hollow cubes, which would float higher.

56. $w = mg$; $w = (0.80 \text{ kg})(9.80 \text{ m/s}^2) = 7.84 \text{ N}$
$F_b = \rho Vg$; $0.54 \text{ N} = (1.0 \times 10^3)\ V\ (0.80 \text{ m/s}^2)$; $V = 5.5 \times 10^{-5} \text{ m}^3$
$\rho = m / V$; $\rho = (0.80 \text{ kg}) / (5.5 \times 10^{-5} \text{ m}^3) = 1.5 \times 10^4 \text{ kg/m}^3$; the density of the crown is less than pure gold, thus the crown is not pure gold.

62. $mg = \rho Vg$; $mg = \rho(Ad)g$
$(2.00 \times 10^3 \text{ kg})(9.80 \text{ m/s}^2) = (1.0 \times 10^3 \text{ kg/m}^3)(1.5 \text{ m})(4.0 \text{ m})\ d\ (9.80 \text{ m/s}^2)$
$d = 0.33 \text{ m}$

67. (d) capillary action depends upon both adhesive forces and surface tension.

72. $h = 2\gamma \cos\theta / \rho gr$; $h = 2(0.073) / (1.0 \times 10^3)(9.80)(1.0 \times 10^{-3}) = 0.015 \text{ m}$

82. the speed of the air between the moving vehicles is greater and hence the pressure is reduced by Bernoulli's equation ; the pressure difference causes inward motion.

92. $Q = V / t$; $V = \text{length} \times \text{width} \times \text{height} = (3.0 \text{ m})(4.5 \text{ m})(6.0 \text{ m}) = 81 \text{ m}^3$
$Q = 81 \text{ m}^3 / 600 \text{ s} = 0.14 \text{ m}^3/\text{s}$

Sample Quiz

Multiple Choice. Choose the correct answer.

__ 1. What is the only modulus present for a liquid ?
A. Young's modulus B. Shear modulus C. Bulk modulus

__ 2. A mass m is hung from identical wires made of aluminum, brass, copper, and steel. Which will stretch the least ?
A. aluminum B. brass C. copper D. steel

__ 3. A glass is filled completly with ice and water. When the ice melts the water level
A. will increase B. will decrease C. will remain the same

__ 4. Which of the following is associated with the law of conservation of energy for fluids ?
A. Archimede's principle
B. Bernoulli's principle
C. Pascal's principle
D. Torrichelli's principle

__ 5. The reason a shower curtain is pulled inward is due to the higher pressure
A. inside the shower B. outside the shower C. none of the choices

__ 6. A block of steel is submerged in the following liquids : alcohol, water, sea water, and blood. In which liquid is the apparent weight the greatest ?
A. alcohol B. blood C. sea water D. water

__ 7. Water flows through a pipe whose diameter is d with a speed v. The diameter of the pipe is halved. By what factor will the speed of the water change if the volume rate of flow remains constant ?
A. (1/4) B. (1/2) C. 2 D. 4

Problems

8. A 6.5-kg mass is hung from a 1.0 mm diameter, 10 m long brass wire. What is the new length of the wire ?

9. A 5.0-cm^3 block of iron is submerged in a container of water.
   A. What is the buoyant force on the block ?
   B. What is the apparent weight of the block in the water ?
   C. What minimum force is required to lift the mass to the surface ?

10. Water flows at a speed of 3.0 m/s in a horizontal pipe 2.0 cm in diameter. The water then enters a horizontal pipe 4.0 cm in diameter.
    A. What is the speed of the water in the large pipe ?
    B. What is the difference in pressure between the two segments ?

CHAPTER 10 Temperature

Chapter Objectives

Upon completion of the unit on temperature, students should be able to :

1. distinguish between heat and temperature.
2. convert temperature measurements between Celsius and Fahrenheit scales, between Celsius and Kelvin scales, and between Fahrenheit and Kelvin scales.
3. state Boyle's law and Charles' law, and apply these laws to compare the temperature, pressure, and volume of an ideal gas.
4. write expressions for, and apply, the perfect (ideal) gas law.
5. explain how absolute zero is determined graphically.
6. write an expression for the thermal expansion (or contraction) of liquids and solids.
7. compare the coefficient of linear expansion, the coefficient of area expansion, and the coefficient of volume expansion.
8. explain the thermal expansion properties of water.
9. determine thermal stress.
10. explain properties of the kinetic theory of gases.
11. write expressions relating the kinetic energy per molecule to the temperature.

Chapter Summary

*__Temperature__ is a relative measure or indication of hotness or coldness. In kinetic theory, temperature is a measure of the average random kinetic energy of the molecules.

*__Heat__ describes energy that is transferred from one object to another because of a temperature difference. The total energy of all molecules of a body or system is the **internal energy**. Heat is internal energy that is added or removed from a body because of a temperature difference relative to another body.

*When heat is transferred between two objects, they are said to be in **thermal contact**. Objects in thermal contact with the same temperature are said to be in **thermal equilibrium**.

*The two most common temperature scales are the **Celsius temperature scale** and the **Fahrenheit temperature scale**, which have ice points of 0°C and 32°F and steam points of 100°C and 212°F.

*The **ideal** or **perfect gas law** gives a relationship for the pressure, volume and temperature of a gas : $pV = NkT$. It describes real, low-density gases fairly well. A gas can be used to measure temperature as a function of pressure at a constant volume, and extrapolation to zero pressure defines absolute zero temperature. **Absolute zero** is the foundation for the **Kelvin temperature scale**, which uses absolute zero and the triple point of water as fixed points.

*The **thermal expansion** of a material is characterized by the coefficients of expansion. The thermal coefficient of linear expansion applies to one dimensional length changes. The thermal coefficient of area expansion (approximately 2α) applies to two-dimensional changes, and the thermal coefficient of volume expansion (approximately 3α) applies to changes in three dimensions. For fluids with no definite shape, only volume expansion is applicable and a special coefficient of volume expansion β is used.

*The **kinetic theory of gases** uses statistical methods to derive the ideal gas law from mechanical principles. The theory shows that the internal energy of an ideal gas is directly proportional to its (absolute) temperature.

*__Diffusion__ is a process of random molecular mixing in which particular molecules move from a region of higher concentration to one of lower concentration. **Osmosis** is the diffusion of a liquid across a permeable membrane because of a concentration gradient.

Important Terms and Relationships

10.1 Temperature and Heat

temperature
heat
internal energy
thermal contact
thermal equilibrium

10.2 The Celsius and Fahrenheit Temperature Scales

Fahrenheit temperature scale : $T_F = (9/5)\,T_C + 32$
Celsius temperature scale : $T_C = (5/9)\,(T_F - 32)$

10.3 Gas Laws and Absolute Temperature

Boyle's law : $p_1V_1 = p_2V_2$
Charles' law : $V_1T_2 = V_2T_1$
ideal (perfect) gas law : $pV = nRT$
absolute zero
Kelvin temperature scale : $T_k = T_c + 273$
kelvins
triple point of water
universal gas constant
Avogadros' number
mole

10.4 Thermal Expansion

thermal coefficient of linear expansion : $\Delta L = L_o\,(1 + \alpha\Delta T)$
thermal coefficient of area expansion : $\Delta A = A_o\,(1 + 2\alpha\Delta T)$
thermal coefficient of volume expansion : $\Delta V = V_o\,(1 + 3\alpha\Delta T)$

10.5 The Kinetic Theory of Gases

kinetic theory of gases : $pV = (1/3)\,Nmv^2$; $(1/2)mv^2 = (3/2)kT$
diffusion
Graham's law : $R_1 / R_2 = [\,m_2 / m_1]^{1/2}$
osmosis

Additional Solved Problems

10.2 The Celsius and Fahrenheit Temperature Scales

Example 1 On a hot day, the temperature reaches 97°F. What is the corresponding temperature on the Fahrenheit scale ?

$T_c = (5/9)\,(T_F - 32)$
$T_c = (5/9)\,(97 - 32) = 36\ °C$

Example 2 The temperature reading on a Celsius thermometer is 125°C. What is the temperature on the Fahrenheit scale ?

$T_F = (9/5)T_C + 32$

$T_F = (9/5)(125) + 32 = 257$ °F

10.3 Gas Laws and Absolute Temperature

Example 1 In a fixed container, the original temperature of a gas is 20°C when the pressure is 1.0 atm. If the pressure is doubled, what is the new temperature ?

given : $T_1 = 293$ K ; $p_1 = 1.0$ atm ; $p_2 = 2.0$ atm

$p_1T_2 = p_2T_1$; make sure the temperatures are in kelvins

$(1.0\text{ atm})(T_2) = (2.0\text{ atm})(293\text{ K})$

$T_2 = 586$ K $T_K = T_C + 273$; $586 = T_C + 273$; $T_C = 313$ °C

Example 2 An ideal gas has an inital volume of 4.0 L , pressure 3.0 atm, and temperature 40°C. What is the new temperature if the volume decreases to 1.5 L and the pressure is increased to 6.0 atm ?

given : $V_1 = 4.0$ L ; $p_1 = 3.0$ atm ; $T_1 = 313$ K ; $V_2 = 1.5$ L ; $p_2 = 6.0$ atm

$p_1V_1T_2 = p_2V_2T_1$

$(3.0\text{ atm})(4.0\text{ L})(T_2) = (6.0\text{ atm})(1.5\text{ L})(313\text{ K})$; $T_2 = 235$ K

Example 3 What is number of moles in 2.0 L of oxygen at a pressure of 3.0 atm and at a temperature of 100°C ?

given : V = 2.0 L ; p = 3.0 atm ; T = 373 K ; R = 0.082 L-atm/mole-K

$pV = nRT$

$(3.0\text{ atm})(2.0\text{ L}) = n\,(0.082\text{ L-atm / mole-K})(373\text{ K})$

$n = 0.20$ mole

$(0.20\text{ mole})(32\text{ g / mole}) = 6.4$ g

10.4 Thermal Expansion

Example 1 A slab of concrete has a length of 15.0 m at a temperature of 0°C. What is the length of the slab if it is heated to a temperature of 40°C ?

given : $L_o = 15.0$ m ; $T_o = 0°C$; $T = 40°C$; $\alpha = 12 \times 10^{-6}\ C^{-1}$

$L = L_o(1 + \alpha\Delta T)$; $L = (15.0\ m)[1 + (12 \times 10^{-6}\ C^{-1})(40\ C°)]$

$L = 15.0072$ m

Example 2 A 500-mL beaker of water is filled to the brim at a temperature of 10°C. How much water will overflow if the water is heated to a temperature of 80°C ?

given : $V_o = 500$ mL ; $T_o = 10°C$; $T = 80°C$; $\beta_{water} = 2.1 \times 10^{-4}\ C^{-1}$;

$\beta_{pyrex} = 9.9 \times 10^{-6}\ C^{-1}$

$\Delta V = V_o[1 + \beta\Delta T]$;

$\Delta V_{water} = (500\ mL)[1 + (2.1 \times 10^{-4}\ C^{-1})(70°C)] = 507.4$ mL

$\Delta V_{pyrex} = (500\ mL)[1 + (9.9 \times 10^{-6}\ C^{-1})(70°C)] = 500.3$ mL

$\Delta V_{overflow} = 7.1$ mL

10.5 The Kinetic Theory of Gases

Example 1 Find the average kinetic energy of the following gases at room temperature.
A. oxygen
B. nitrogen

given : $T = 293$ K ; $k = 1.38 \times 10^{-23}$ J/K

A. $K = (3/2)\,kT$; $K = (3/2)(1.38 \times 10^{-23}\ J/K)(293\ K) = 6.07 \times 10^{-21}$ J

B. same, since the temperature is the same

Example 2 Find the rms speed of an oxygen molecule at room temperature. At what temperature does the speed of the molecule double ?

given : $T = 20°C$; $k = 1.38 \times 10^{-23}$ J/K

$K = (3/2)kT$; $K = (3/2)(1.38 \times 10^{-23} J/K)(293\ K) = 6.07 \times 10^{-21}\ J$
$K = (1/2)\ mv^2$; $m = (32\ u)\ (1.66 \times 10^{-27}\ kg/u) = 5.3 \times 10^{-26}\ kg$
$(6.07 \times 10^{-21}\ J) = (1/2)(5.3 \times 10^{-26}\ kg)\ v^2$; $v = 479\ m/s$

Solutions to paired problems and selected problems

6. $T_C = (5/9)\ (T_F - 32)$
   (a) $T_C = (5/9)\ (0 - 32) = -18°C$
   (b) $T_C = (5/9)\ (1500 - 32) = 816°C$
   (c) $T_C = (5/9)\ [\ (-20) - 32\ ] = -29°C$
   (d) $T_C = (5/9)\ [\ (-40) - 32\ ] = -40°C$; this is the temperature where the the Celsius and Fahrenheit temperatures are equal.

11. $T_F = (9/5)\ T_C + 32$
    $T_F = (9/5)(38.8) + 32 = 102°F$

15. (a) if the pressure is constant and the temperature is increased, the volume must increase ; since the volume increases, the density decreases.
    (b) if the volume is held constant, the mass is also constant, therefore the density must also remain constant.

20. $T_K = T_C + 273$
    (a) $T_K = 0 + 273 = 273\ K$
    (b) $T_K = 100 + 273 = 373\ K$
    (c) $T_K = 20 + 273 = 293\ K$
    (d) $T_K = -35 + 273 = 238\ K$

24. (a) H_2O : gram molecular weight is 2 g (hydrogen) + 16 g (oxygen) = 18 g
    (40 g) (1 mole / 18 g) = 2.2 moles
    (b) H_2SO_4 : gram molecular weight is 2 g (H) + 32 g (S) + 64 g(oxygen) = 98 g
    (245 g) (1 mole / 98 g) = 2.5 moles

(c) NO_2 : 7 g (nitrogen) + 32 g (oxygen) = 39 g
(117 g) (1 mole / 39 g) = 3.0 moles
(d) SO_2 is a gas
56 L (mole / 22.4 L) = 2.5 moles

31. $pV = NkT$
$(2.53 \times 10^5 \text{ Pa})(5.0 \text{ L}) = N (1.38 \times 10^{-23} \text{ J/K})(293 \text{ K})$; $N = 3.1 \times 10^{23}$ molecules

an alternative solution :
$pV = nRT$
(2.5 atm)(5.0 L) = n (0.082 L-atm / mole-K) (293 K) ; n = 0.52 mole
$(0.52 \text{ mole})(6.02 \times 10^{23} \text{ molecules / mole}) = 3.1 \times 10^{23}$ molecules

37. (c)

44. (a) $L = L_o (1 + \alpha\Delta T)$; $L = (60 \text{ cm}) [1 + (17 \times 10^{-6} \text{ C}^{-1})(65°\text{C}) = 60.07 \text{ cm}$
(b) $A_o = \pi r^2$; $A_o = (3.14)(0.750 \text{ cm})^2 = 1.77 \text{ cm}^2$
$\Delta A = A_o (1 + 2 \alpha\Delta T) = (1.77 \text{ cm}^2) [1 + 2 (17 \times 10^{-6} \text{ C}^{-1})(65)$
$\Delta A = 3.91 \times 10^{-3} \text{ cm}^2$

50. $\rho = m / V$; the mass remains constant, therefore set up a relationship between the density and the volume :
$\rho_1 V_1 = \rho_2 V_2$; $V_2 = V_1 [1 + \beta\Delta T]$
$\rho_1 V_1 = \rho_2 V_2$; $13.6 \text{ g/cm}^3 (V_1) = (\rho_2) V_1 [1 + (1.8 \times 10^{-4} \text{ C°}^{-1})(80 \text{ C°})]$
$\rho_2 = 13.4 \text{ g/cm}^3$

55. $K = (3/2) kT$
the Celsius temperature doubles, but this is not absolute temperature.
$T_1 = 20 + 273 = 293$ K ; $T_2 = 40 + 273 = 313$ K
$K_2 / K_1 = T_2 / T_1$; $K_2 / K_1 = 313 \text{ K} / 293 \text{ K} = 1.07$
the answer is d

62. from the Exercise 55. : $K_2 / K_1 = T_2 / T_1$
$2 K_1 / K_1 = T_2 / (298 \text{ K})$; $T_2 = 598$ K
$T_c = 598 \text{ K} - 273 = 323$ °C

68. first let L_1 be the length when it is heated :

$L_1 = L_o\ (1 + \alpha\Delta T)$; $L_1 = (1.0\ m)\ [\ 1 + (24 \times 10^{-6}\ C^{-1})(20\ °C)] = 1.00048\ cm$

$L_2 = (1.0\ m)\ [\ 1 + (24 \times 10^{-6}\ C^{-1})(-20°C)] = 0.99999\ cm$

Chapter 10

Sample Quiz

Multiple Choice. Choose the best answer.

__ 1. Which of the following is the greatest temperature ?
A. 0°F B. 0°C C. 263 K D. -5°C

__ 2. A container of water is heated from 0°C to 4°C. What happens to the density of the water ?
A. decreases B. increases C. remains the same

__ 3. A container of water is heated from 10°C to 20°C. What happens to the density of the water ?
A. decreases B. increases C. remains the same

__ 4. The temperature of a gas is doubled while the pressure remains constant. By what factor does the pressure change ?
A. (1/4) B. 1/2 C. 2 D. 4

__ 5. The pressure of a gas doubles while the volume doubles. By what factor does the temperature change ?
A. (1/4) B. (1/2) C. 2 D. 4

__ 6. A fixed container holds oxygen and hydrogen gases at the same temperature. Which of the following expressions is correct ?
A. The oxygen molecules have the greater kinetic energy.
B. The hydrogen molecules have the greater kinetic energy.
C. The oxygen molecules have the greater speed.
D. The hydrogen molecules have the greater speed.

__ 7. The average speed of the molecules of a gas is doubled. By what factor does the temperature change ?
A. (1/4) B. (1/2) C. 2 D. 4

Problems

8. An ideal gas has a pressure of 2.5 atm, a volume of 1.0 L and a temperature of 30°C.

   A. If the gas is oxygen, how much mass does the gas have ?

   B. If the pressure doubles and the volume quadruples, what is the new temperature in degrees Celsius ?

9. The average speed of a oxygen molecule is 600 m/s.

   A. What is the kinetic energy of the molecule ?

   B. What is the temperature of the gas ?

10. A steel girder is 20.0 m long at a temperature of -18°C. What is its length on a hot day when the temperature is 38°C ?

CHAPTER 11 Heat

Chapter Objectives

Upon completion of the unit on heat, students should be able to :

1. define heat in terms of energy transfer.
2. define a calorie and a kilocalorie.
3. distinguish between a Calorie and a calorie.
4. write and apply the expression for intrinsic energy value in terms of the heat of combustion.
5. define terms such as: specific heat, latent heat, heat of fusion, heat of vaporization, sublimation, melting point, freezing point, boiling point, and condensation point.
6. write and apply heat exchange expressions involving specific heat and latent heat.
7. interpret a phase diagram for a substance.
8. interpret a temperature versus heat diagram for a substance.
9. explain, write, and apply expressions for the three types of heat transfer.
10. explain relative humidity and the process of evaporation.
11. explain how a change in pressure can affect the phases of matter at various temperatures.

Chapter Summary

*__Heat__ is energy that is transferred from one object or system to another because of a temperature difference. Heat is energy in transit and is expressed with the joule (J) energy unit.

*Other units commonly used to measure heat are the calorie and the kilocalorie. A kilocalorie is the amount of heat required to raise the temperature of 1 kilogram of water by 1C°. (The kilocalorie is the food Calorie.)

*The **heat of combustion** is a measure of the intrinsic energy values of substances, mainly food and fuels.

*The relationship between heat units and the standard (mechanical) energy unit, the joule is called the **mechanical equivalent of heat** : 1 kcal = 4.19×10^3 J (1 cal = 4.19 J).

*__Specific heat__ is the amount of heat required to raise the temperature of 1 kilogram of a substance by 1C°. For water the specific heat is 1.0 kcal/kg-C°.

*Matter normally exists in three phases: solid, liquid, and gas. **Latent heat** is the heat involved in a phase change and does not go into changing the temperature, but into the work of changing the phase.

*Three mechanisms of heat transfer are **conduction**, **convection**, and **radiation**. The conducting ability of a material is characterized by its **thermal conductivity**. Convection involves mass transfer; conduction does not. Radiation is the transfer of energy by electromagnetic waves and requires no transfer medium. A good emitter of radiation is also a good absorber.

*Evaporation is a cooling process because energy is required to bring about the phase change.

*The vapor pressure of the water in air is expressed in terms of the **relative humidity**, which is the ratio of the partial pressure of the moisture to the saturated vapor pressure at a given temperature, or the ratio of the actual moisture content to the maximum moisture content at a given temperature. The relative humidity is 100% at the **dew point**.

Important Terms and Relationships

11.1 Units of Heat

heat
calorie (cal)
kilocalorie (kcal)
British thermal unit (Btu)
mechanical equivalent of heat : $1 \text{ kcal} = 4.19 \times 10^3 \text{ J}$
heat of combustion : $Q = mH$

11.2 Specific Heat

specific heat (capacity) : $Q = mc\Delta t$

11.3 Phase Changes and Latent Heat

solid phase
melting point
freezing point

liquid phase
gaseous (vapor) phase
boiling point
condensation point
sublimation
latent heat : $Q = mL$
latent heat of fusion (water : 3.3×10^5 J/kg or 80 kcal/kg)
latent heat of vaporization (water : 22.6×10^5 J/kg or 540 kcal/kg)
phase diagrams

11.4 Heat Transfer

conduction : $\Delta Q/\Delta t = kA\Delta T / d$
radiation
infrared radiation
Stefan's law $P = \sigma A e T^4$
Radiation power or loss : $P_{net} = \sigma A e(T_s^4 - T^4)$
emissivity
black body
intensity : $I = P_{net} / A = \sigma\, e(T_s^4 - T^4)$

11.5 Evaporation and Relative Humidity

saturated vapor pressure
relative humidity
dew point

Additional Solved Problems

11.1 Units of Heat

Example 1 A person eats a package of candy which is 240 Cal. How many times will the person need to lift a 10 kg mass a height of 2.0 m to equal the candy's intrinsic energy ?

given : Q = 240 Cal or 240 kcal ; m = 10 kg ; h = 2.0 m

$Q = 240 \text{ kcal} (4190 \text{ J / kcal}) = 1.01 \times 10^6$ J

to lift the mass once : $U = mgh = (10 \text{ kg})(9.80 \text{ m/s}^2)(2.0 \text{ m}) = 196$ J

times = $(1.01 \times 10^6 \text{ J}) / (196 \text{ J}) = 5.15 \times 10^3$ times

Example 2 How much heat will be released upon the combustion of 1.2 kg of alcohol ?

given : $m = 1.2$ kg ; $H = 2.7 \times 10^7$ J/kg

$Q = mH$; $Q = (1.2 \text{ kg})(2.7 \times 10^7 \text{ J/kg}) = 3.2 \times 10^7$ J

11.2 Specific Heat

Example 1 A 1.0 kg copper container holds 0.30 kg of water at room temperature. How much heat must be added to raise the temperature to 80°C ?

water : $m = 0.30$ kg ; $c = 1.0$ kcal/kg-C° ; $\Delta T = 60$ C° (room temp 20°C)

copper : $m = 1.0$ kg ; $c = 0.093$ kcal/kg-°C

Q_1: heat copper ; Q_2 : heat water

$Q = Q_1 + Q_2$

$Q = mc\Delta T + mc\Delta T$

$Q = (1.0 \text{ kg})(0.093 \text{ kcal/kg-C°})(60 \text{ C°}) + (0.30 \text{ kg})(1.0 \text{ kcal/kg-C°})(60 \text{ C°})$

$Q = 23.6$ kcal

Example 2 A 0.50-kg unknown substance is heated to a temperature of 150°C and is then placed into 0.080 kg of water initially at 10°C. It is observed when the mass is placed into the water, they reach a thermal equilibrium of 25°C. What is the specific heat of the substance ?

given : unknown -- $m = 0.50$ kg ; $\Delta T = 125$ C°

water -- $m = 0.080$ kg ; $c = 1.00$ kcal/kg-°C ; $\Delta T = 15$ C°

$Q_{lost} = Q_{gained}$

$(mc\Delta T)_{unknown} = (mc\Delta T)_{water}$

$(0.50 \text{ kg})(c)(125 \text{ C°}) = (0.080 \text{ kg})(1.0 \text{ kcal / kg-C°})(15 \text{ C°})$

$c = 1.9 \times 10^{-2}$ kcal / kg-C°

Example 3 A 100 g aluminum calorimeter cup contains 200 g of water at a temperature of 10°C. A beaker containing 100 mL of water at 80°C is poured into the calorimeter. What is the final temperature of the mixture ?

given : aluminum -- $m = 0.100$ kg ; $c = 0.22$ kcal/kg-C° ; $T_o = 10$°C

water (cold) -- $m = 0.200$ kg ; $c = 1.00$ kcal/kg-C° ; $T_o = 10$°C

water (hot) -- $m = 0.100$ kg ; $c = 1.00$ kcal/kg-C° ; $T_o = 80$°C

$Q_{gained} = Q_{lost}$
$(mc\Delta T)_{al} + (mc\Delta T)_{water} = (mc\Delta T)_{water}$
$(0.100 \text{ kg})(0.22 \text{ kcal / kg-C°})(T - 10) + (0.200)(1.00 \text{ kcal / kg-C°})(T - 10) =$
$(0.100)(1.0 \text{ kcal / kg-C°})(80 - T)$
$(0.022)T - 0.22 + (0.20)T - 2.0 = 8.0 - (0.10)T$
$0.322\,T = 10.22$; $T = 31.7°C$

11.3 Phase Changes and Latent Heat

Example 1 How much heat is needed to change 0.50 kg of ice at -30°C to steam at a temperature of 150°C ?

given : $m = 0.50$ kg ; $T_o = -30°C$; $T = 150°C$
Q_1 : heat ice from -30°C to 0°C ; Q_2 : melt ice ;
Q_3 : heat water from 0°C to 100°C ; Q_4 : change water to steam
Q_5 : warm steam from 100°C to 150°C
$Q = Q_1 + Q_2 + Q_3 + Q_4 + Q_5$
$Q = mc\Delta T + mL_f + mc\Delta T + mL_v + mc\Delta T$
$Q = (0.50 \text{ kg})(0.50 \text{ kcal / kg-C°})(30 \text{ C°}) + (0.50 \text{ kg})(80 \text{ kcal/kg}) +$
$(0.50 \text{ kg})(1.0 \text{ kcal / kg-C°})(100 \text{ C°}) + (0.50 \text{ kg})(540 \text{ kcal/kg}) +$
$(0.50 \text{ kg})(0.43 \text{ kcal / kg-C°})(50 \text{ C°}) = 378 \text{ kcal}$

Example 2 A large storage tank holds 50 kg of water at 30°C.
A. How much water at 0°C must be added to reduce the temperature to of the water in the tank to 10°C ?
B. How much ice at 0°C must be added to reduce the temperature to of the water in the tank to 10°C ?

given : $m = 50$ kg ; $T_o = 30°C$

A. $Q_{lost} = mc\Delta T$; $Q_{lost} = (50 \text{ kg})(1.0 \text{ kcal / kg-C°})(20 \text{ C°}) = 1.0 \times 10^3 \text{ kcal}$

$Q_{gained} = mc\Delta T$; $1.0 \times 10^3 \text{ kcal} = m(1.0 \text{ kcal / kg-C°})(10 \text{ C°})$;

$m = 1.0 \times 10^2$ kg

B. $Q_{gained} = mL_f + mc\Delta T$

$1.0 \times 10^3 \text{ kcal} = m(80 \text{ kcal/kg}) + m(1.0 \text{ kcal / kg-C°})(10 \text{ C°})$; $m = 11$ kg

Example 3 A beaker holds 100 g of water at a temperature of 70°C. Twenty grams of ice at 0°C are added. What is the final temperature and contents of the final mixture ?

given : water -- m = 0.100 kg ; T_o = 70°C

ice -- m = 0.020 kg ; T_o = 0°C

$Q_{lost} = Q_{gained}$

(0.100 kg)(1.0 kcal / kg-C°) (70 - T) =

(0.020 kg)(80 kcal / kg) + (0.020) (1.0 kcal / kg-C°) (T - 0)

7.0 - (0.1) T = 1.6 + 0.02 T

5.4 = 0.12 T ; T = 45 °C ; all water

11.4 Heat Transfer

Example 1 A window 0.50 cm thick has dimensions of 1.5 m by 1.5 m. What is the rate of heat conduction through the window if the outside temperature is 38°C and the inside temperature 22°C ? How much heat flows through the window in 5.0 h ?

given : k_{glass} = 0.084 J/m-s-C° ; A = (1.5 m)(1.5 m) = 2.25 m^2 ; ΔT = 16 C°

d = 5.0×10^{-3} m

$\Delta Q / \Delta t = kA (\Delta T) / d$

$\Delta Q / \Delta t = (0.084 \text{ J / m-s C°})(2.25 \text{ m}^2)(16 \text{ C°}) / (0.50 \times 10^{-2} \text{ m})$

$\Delta Q / \Delta t = 6.0 \times 10^2$ J/ s

$\Delta Q / (1.8 \times 10^4 \text{ s}) = (6.0 \times 10^2 \text{ J / s})$; $\Delta Q = 1.1 \times 10^7$ J

Example 2 A radiating body has an emissivity of 0.50 and its exposed area is 0.80 m^2. The temperature of the body is 100°C and the surrounding temperature is 20°C. What is rate heat is radiated from the body ?

$P_{net} = \sigma A e (T_s^4 - T^4)$

$P_{net} = (5.67 \times 10^{-8} \text{ W / m}^2\text{-K}^4)(0.80 \text{ m}^2)(0.50)[(373 \text{ K})^4 - (293 \text{ K})^4]$

P_{net} = 272 W

Solutions to paired and other selected problems

4. gasoline has a greater intrinsic heat value than alcohol therefore it has a greater intrinsic heat value than gasohol.

12. first find the volume in m^3 ; $(1 \text{ gal})(1 \text{ m}^3 / 264 \text{ gal}) = 3.79 \times 10^{-3} \text{ m}^3$
next find the mass : $\rho = m / V$; $(0.68 \times 10^3 \text{ kg/m}^3) = m / (3.79 \times 10^{-3} \text{ m}^3)$
$m = 2.6 \text{ kg}$
now find the energy from combustion :
$$Q = mH = (2.6 \text{ kg})(4.8 \times 10^7 \text{ J/kg}) = 1.2 \times 10^8 \text{ J}$$

16. three liters of water have the mass of 3.0 kg
$Q = mc\Delta T$; $12 \text{ kcal} = (3.0 \text{ kg})(1.0 \text{ kcal/kg-C}^\circ)\ \Delta T$; $\Delta T = 4 \text{ C}^\circ$; (c)

22. $Q_{lost} = Q_{gained}$
$mc\Delta T_{water} = mc\Delta T_{alcohol}$
$(1.0 \text{ kg})(1.0 \text{ kcal/kg-C}^\circ)(40\ ^\circ\text{C} - T) = (0.79 \text{ kg})(0.58 \text{ kcal/kg-C}^\circ)(T - 20\ ^\circ\text{C})$
$40 - T = 0.46\,T - 9.2$; $T = 34^\circ\text{C}$

28. $Q_{lost} = Q_{gained}$
$(mc\Delta T)_{aluminum} = (mc\Delta T)_{water}$
$(0.35 \text{ kg})(0.22 \text{ kcal/kg-C}^\circ)(100\ ^\circ\text{C} - T) = (0.050 \text{ kg})(1.0 \text{ kcal/kg-C}^\circ)(T - 10^\circ\text{C})$
$7.7 - (0.077)T = (0.050)\,T - (0.50)$
$T = 65^\circ\text{C}$

32. $Q = mc\Delta T + mL_f$
$Q = (0.030 \text{ kg})(130 \text{ J/kg-C}^\circ)(308 \text{ C}^\circ) + (0.030 \text{ kg})(0.25 \times 10^5 \text{ J/kg}) = 1.95 \times 10^3 \text{ J}$
$K = Q = (1/2)\,mv^2$
$1.95 \times 10^3 \text{ J} = (1/2)(0.030 \text{ kg})(v^2)$; $v = 4.0 \times 10^2 \text{ m/s}$

37. no ; specific heat only deals with when there is no phase change ; when there is a phase change, latent heats must be used.

42. Q_1: cool water ; Q_2 = freeze water ; Q_3 = reduce temperature of ice

$Q = Q_1 + Q_2 + Q_3$

Q = (2.0 kg)(1.0 kcal/kg-C°)(20 C°) + (2.0 kg)(80 kcal/kg) +
(2.0 kg)((0.50 kcal/kg-C°)(10 C°) = 2.1 x 10^2 kcal

46. Q_1 : warm ice to 0°C ; Q_2 : cool tea ; Q_3 : melt ice

$Q_1 = mc\Delta T$ = (0.20 kg)(0.50 kcal/kg-C°)(10 C°) = 1.0 kcal

$Q_2 = mc\Delta T$ = (0.10 kg)(1.0 kcal/kg-C°)(50 C°) = 5.0 kcal

$Q_{lost} = Q_{gained}$

$Q_2 = Q_1 + Q_3$

5.0 kcal = 1.0 kcal + Q_3 ; Q_3 = 4.0 kcal

$Q_3 = mL_f$; 4.0 kcal = m (80 kcal / kg) ; m = 0.050 kg (mass if ice that melts)

therefore the mass of water is 0.100 kg + 0.050 kg = 0.15 kg

50. Q_1 : cool water to freezing point ; Q_2 : freeze water ; Q_3 : cool ice

$Q = Q_1 + Q_2 + Q_3$

$Q = mc\Delta T + mL_f + mc\Delta T$

Q = (0.50 kg)(1.0 kcal/kg-C°)(20 C°) + (0.50 kg)(80 kcal/kg) +
(0.50 kg)(0.50 kcal/kg-C°)(5.0 C°) = 51 kcal

57. Bridges have very little insulation, and so react to changes in air temperature quickly. The bridge is exposed in all areas whereas a road is exposed only on one side.

66. (a) the greater the R-value, the greater the insulation value.
(b) L = k (R-value)
(1) L = (0.42)(10) = 4.2 in.
(2) L = (5.1) (10) = 51 in.

70. $\Delta T = (\Delta Q/\Delta t)(d/k)(1/A) = HR / A$; $H = \Delta Q/\Delta T$; $R = d/k$

for steady state (H/A) is the same for all layers

for each layer : $T_4 - T_3 = (H/A)R_1$; $T_3 - T_2 = (H/A)R_2$; $T_2 - T_1 = (H/A)R_3$

and $T_4 - T_1 = (H/A)(R_1 + R_2 + R_3)$

$\Delta T = (\Delta Q/\Delta t)\ (1/A)[\ (d_1/R_1) + (d_2/R_2) + d_3/R_3)]$

$\Delta Q / \Delta T = A\Delta T / [\ (d_1/k_1) + (d_2/k_2) + (d_3/k_3)]$

$\Delta Q/\Delta T$ = (17.5 m^2)(25 C°)/[(0.020 m /0.059 J/m-s-C°) + (0.15/1.3) + (0.070/0.71)]

$\Delta Q/\Delta T = 1.2 \times 10^2$ J/s
$\Delta Q = (1.2 \times 10^2 \text{ J/s})(3600 \text{ s}) = 4.3 \times 10^5$ J

74. $A = 4\pi r^2$; $A = 4\pi(1.5 \times 10^{11} \text{ m})^2 = 2.8 \times 10^{23} \text{ m}^2$
$E/s = IA$; $E/s = (0.33 \text{ kcal/m}^2\text{-s})(2.8 \times 10^{23} \text{ m}^2) = 9.2 \times 10^{22}$ kcal/s

81. Fogs form due to the cooler temperatures found in the valleys. This is because of cooling due to radiation losses at night and such fogs are often called radiation fogs. The Sun increases the air temperature above the dew point and the fog dissipates.

84. $Q/m = c\Delta T + L_v = (1.0 \text{ kcal/kg-C°}) + 540 \text{ kcal} = 605$ kcal
from section 11.5 (Evaporation and Relative humidity)
1 kcal = m (540 kcal/kg) ; Q / m = 605 kcal/kg
(605 kcal / kg) m = Q
(605 kcal / kg) m = 1.0 kcal
m = 1.7 g or 1.7 mL

93. $U = mgh = m(9.80 \text{ m/s}^2)(75 \text{ m}) = (7.35 \times 10^2)$ m
$Q = U = mc\Delta T$; (7.35×10^2) m = m (4190 J/kg-C°)(ΔT) ; $\Delta T = 0.18$ C°

98. Q_1 = heat ice to freezing point ; Q_2 = melt ice ; Q_3 = heat water to 20°C
$Q = Q_1 + Q_2 + Q_3$
$Q = mc\Delta T + mL + mc\Delta T$
Q = (0.75 kg)(0.50 kcal / kg-C°)(10 C°) + (0.75 kg)(80 kcal/kg) +
(0.75 kg)(1.0 kcal / kg-C°)(20 C°) = 79 kcal

Sample Quiz

Multiple Choice. Choose the correct answer.

__ 1. Equal masses of copper, aluminum, lead, and steel at 100°C are each added to a container with 1.0 L of water. Which metal would produce the greatest temperature change in water ?
A. aluminum B. copper C. lead D. steel

__ 2. The reason ocean temperatures do not vary drastically is that
A. water has a high rate of conduction
B. water is a good radiator
C. water has a high specific heat
D. water has a low specific heat

__ 3. Equal massess of water at 20°C and 80°C are mixed. What is the final temperature of the mixture ?
A. 30°C B. 40°C C. 50°C D. 60°C

__ 4. Equal masses of ice at 0°C and steam at 100°C are mixed. What is the final temperature of the mixture ?
A. 0°C B. 50°C C. 100°C D. 280°C

__ 5. Which of the following is the best conductor of heat ?
A. aluminum B. copper C. steel D. silver

__ 6. The temperature of a radiating body is doubled. By what factor is the rate of radiation changed ?
A. 2 B. 4 C. 8 D. 16

__ 7. Which of the following humidities has the greastest moisture content per unit volume ?
A. 20% B. 40% C. 60% D. 80%

Problems

8. How much heat must be removed from 0.20 kg of steam at 130°C to change it completely to water at 20°C ?

9. An 200-g aluminum calorimeter holds 300 g of water at 75°C. Twenty grams of ice at 0°C are added to the water. What is the final temperature of the mixture ?

10. A 5000-W heater is used to heat water. How long will it take the heater to heat 20 kg of water from 10°C to 50°C ? (assume no heat loss)

CHAPTER 12 Thermodynamics

Chapter Objectives

Upon completion of the unit on thermodynamics, students should be able to :

1. distinguish between an open and a closed system.
2. describe a heat reservoir.
3. describe an irreversible process and a reversible process.
4. apply the first law of thermodynamics for different processes including :
   (a) an isobaric process
   (b) an adiabatic process
   (c) an isothermal process
   (d) an isometric process
5. find the area of a p-V diagram to determine the work for a process.
6. sketch a p-V , p-T, and a V-T diagram for reversible processes.
7. state the second law of thermodynamics in the following ways :
   (a) heats flows from a warm body to a colder body.
   (b) heat cannot be transformed completely into mechanical work.
   (c) it is impossible to construct a perpetual motion machine.
8. descibe entropy and find the net entropy for an isolated system, for a total system.
9. describe the concept of the "heat death of the universe."
10. describe the heat engine, and calculate the thermal efficiency for a heat engine.
11. describe the operation of a heat pump.
12. calculate the Carnot efficiency for an ideal engine.
13. describe the implications of the third law of thermodynamics.

Chapter Summary

*__Thermodynamics__ deals primarily with the transfer or action (dynamics) of heat.

*A **system** is a definite quantity of matter enclosed by boundaries, real or imaginary. An open system or closed system refers to whether or not mass can be transferred into or out of a system. There is no heat transfer in or out of a thermally isolated system. A completely isolated system has no interaction with its surroundings.

*A **heat reservoir** is a system with an unlimited heat capacity for which the addition or removal of heat causes no temperature change.

*A **thermodynamic equation of state** is a mathematical relationship of the thermodynamic or state variables. The equation for a quantity of ideal gas is $pV = NkT$, where the variables or thermodynamic coordinates are (p , V, T). A

process is a change in the state or thermodynamic coordinates of a system. A **reversible process** is one with a known process path. The initial and final states of a process are not known refers to an **irreversible process**.

The **first law of thermodynamics** is a statement of the conservation of energy applied to thermodynamic systems: $Q = \Delta U + W$. (+Q and +W when heat is added to or work is done by a system ; -Q and -W when heat is removed or work is done on a system.

* Iso-processes are ones in which one of the thermodynamic variables is held constant. For a quantity of gas, these are: **isobaric** -- constant pressure, **isometric**-- constant volume, and **isothermal** -- constant temperature, In an **adiabatic** process, there is no heat transfer for a system - constant Q. On a p-V graph, the work is the area under the process path.

*The **second law of thermodynamics** specifies the direction in which a process can naturally take place. Different statements are used to describe the second law according to the situation or application, e.g.,

Heat will not flow spontaneously from a colder body to a warmer body.
Heat energy cannot be transformed completely into mechanical work (and vice versa).
It is impossible to construct an operational perpetual-motion machine.

* **Entropy** is a function of the state of a system, and the change in entropy (ΔS) gives the spontaneous or natural direction of a process. The entropy of an isolated system increases in every natural process. As a second law statement : the total entropy of the universe increases in every natural process.

*The entropy of a system may decrease, but only if this is the result of a greater increase in entropy, such as from work done on the system, so there is an overall net increase in entropy. An increase in entropy of a system can be viewed as corresponding to a loss in the system's ability to do useful work. Entropy is sometimes viewed as a measure of the disorder of a system, with disorder and entropy being directly related -- the greater the disorder, the greater the entropy.

*A **heat engine** is any device that converts heat to work. The **thermal efficiency** of a heat engine is the ratio of its work output and its heat input. A statement of the second law of thermodynamics applied to heat engines is : a heat engine operating in a cycle cannot convert its heat input completely into work.

* A **heat pump** is a device that transfers energy from a low-temperature reservoir to a high-temperature reservoir. This requires work input , and a statement of the second

law of thermodynamics applied to heat pumps is : no heat pump operating in a cycle can transfer heat energy from a low-temperature reservoir to a high-temperature reservoir without the application of work.

*The ideal **Carnot cycle** for a heat engine consists of two isotherms and two adiabats. This is the most efficient cycle, and the **Carnot efficiency** gives an unattainable, upper limit of efficiency.

*The **third law of thermodynamics** states that it is impossible to obtain a temperature of absolute zero.

Important Terms and Relationships

12.1 Thermodynamic Systems, States, and Processes

thermodynamics
system
open system
closed system
thermally isolated system
completely isolated system
heat reservior
equation of state
process
irreversable process
reversable process

12.2 The First Law of Thermodynamics

first law of thermodynamics : $Q = \Delta U + W$
work done by a gas (constant pressure) $W = p\Delta V$
isobaric process
isobar
isometric process
isomet
adiabatic process
adiabat

12.3 The Second Law of Thermodynamics and Entropy

second law of thermodynamics
entropy : $\Delta S = Q / T$
isentropic process

12.4 Heat Engines and Heat Pumps

heat engine
Otto cycle
efficiency : $\varepsilon = W / Q_{in} = (Q_{hot} - Q_{cold}) / Q_{hot} = 1 - (Q_{cold} / Q_{hot})$
heat pumps
coefficient of performance : $cop = Q_{cold} / W_{in} = Q_{cold} / (Q_{hot} - Q_{cold})$

12.5 The Carnot Cycle and Ideal Heat Engines

Carnot cycle
Carnot (ideal) efficiency : $\varepsilon = 1 - (T_{cold} / T_{hot}) = (T_h - T_c) / T_h$
third law of thermodynamics

Additional Solved Problems

12.2 The First Law of Thermodynamics

Example 1 An ideal gas undergoes an adiabatic expansion in which the work done by the gas is 4.0×10^3 J.
A. What is the change in heat ?
B. What is the change in internal energy ?
C. Will the temperature increase, decrease, or remain the same ?

given : $W = 4.0 \times 10^3$ J ; adiabatic process means $Q = 0$

(a) since the process is adiabatic, the change in heat is zero 0
(b) $Q = \Delta U + W$
$0 = \Delta U + (4.0 \times 10^3 \text{ J})$; $\Delta U = -4.0 \times 10^3$ J
(c) since the change in internal energy is (-), the temperature decreases

Example 2 Under a constant pressure of 2.0 kPa, the volume of a gas is increased from 1.5 m^3 to 4.0 m^3 while the amount of heat added is 2.00 kcal.
A. What is the amount of work done ?
B. What is the change in internal energy ?

given : $p = 2.0$ kPa $= 2.0 \times 10^3$ Pa ; $V_o = 1.5\ m^3$; $V = 4.0\ m^3$

$Q = 200$ kcal $= 8.38 \times 10^3$ J

(a) $W = p\,\Delta V$; $W = (2.0 \times 10^3\ Pa)(2.5\ m^3) = 5.0 \times 10^3\ J$

(b) $Q = \Delta U + W$; $8.38 \times 10^3\ J = \Delta U + 5.0 \times 10^3\ J$; $\Delta U = 3.38 \times 10^3\ J$

Example 3 During an isothermal process, 5.0 kcal of heat are removed from a system.
A. What is the change in internal energy ?
B. What is the work done in the process ?

given : isothermal process --- $\Delta U = 0$; $Q = -5.0$ kcal $= -2.1 \times 10^4$ J

(a) since the process is isothermal there is no change in internal energy.
(b) $Q = \Delta U + W$
$-2.1 \times 10^4\ J = 0 + W$; $W = -2.1 \times 10^4\ J$

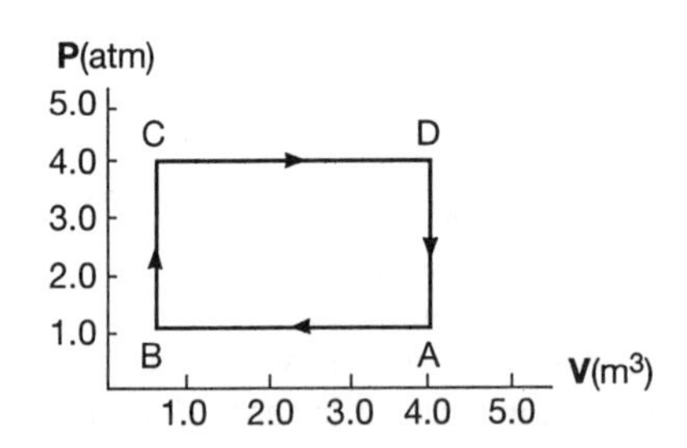

Fig. 12.1

Example 4 In Fig. 12.1, an ideal gas beginning at point A follows the process path ABCDA.
A. What is the amount of work done for a complete cycle ?
B. What is the change in internal energy for a complete cycle ?
C. What is the change in heat for a complete cycle ?
D. At what point is the temperature the greatest ?

(a) the work done for a complete cycle equals the area for the rectangle :

$W = bh = (3.0\ m^3)(3.0 \times 10^5\ N/m^2) = 9.0 \times 10^5\ J$

(b) for a complete cycle, the product of pV is the same in comparison to the beginnning ; the temperature is proportional to the product of pV, there is no change in internal energy.

(c) $Q = \Delta U + W$

$Q = 0 + 9.0 \times 10^5\ J$; $Q = 9.0 \times 10^5\ J$

(d) the temperature is the greatest where the product of pV is the greatest - point D

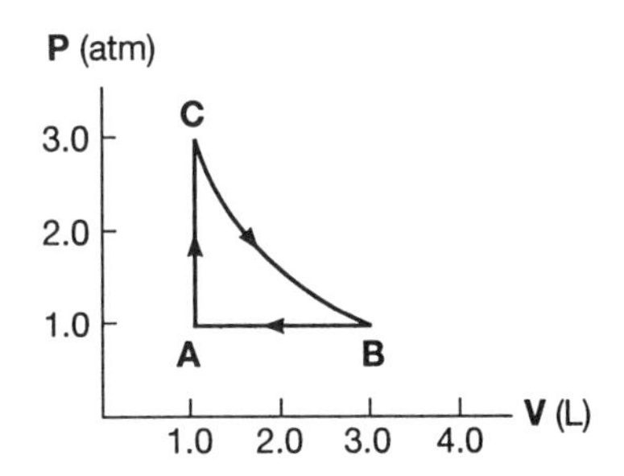

Fig. 12.2

Example 5 One mole of an ideal gas undergoes a process BACB as shown in Fig.12.2. The heat added from C to B is 4.0×10^2 J.

A. What is the temperature of the gas at point B ?
B. What is the tempertaure of the gas at point A ?
C. What work is done for the process BA ?
D. What work is done for a complete cycle ?

given : n = 1.0 mole ; $Q_{CB} = 4.0 \times 10^2\ J$

(a) $pV = nRT$; at point B :

(1.0 atm)(3.0 L) = (1 mole)(0.082 L-atm/mole-K)(T)

T = 365 K

(b) at point A : (1.0 atm)(1.0 L) = (1.0 mole)(0.082 L-atm / mole-K) T

T = 122 K

(c) $W = p\Delta V$; $W = (1.0 \times 10^5\ N/m^2)(-2.0 \times 10^{-3}\ m^3) = -2.0 \times 10^2\ J$

(d) $W_{AC} = 0$; $W_{CB} = 4.0 \times 10^2\ J$; $W_{total} = W_{BA} + W_{AC} + W_{CB} = 200\ J$

12.3 The Second Law of Thermodynamics

Example 1 Find the change in entropy when 1.5 kg of steam at 100°C are condensed to water at 100°C.

given : $m = 1.5$ kg ; $\Delta T = 0$ C°

first find the heat : $Q = mL$; $Q = -(1.5 \text{ kg})(22.6 \times 10^5 \text{ J/kg}) = -3.39 \times 10^6$ J

$\Delta S = Q / T$; $\Delta S = (-3.39 \times 10^6 \text{ J}) / 373 \text{ K} = -9.1 \times 10^3$ J/K

Example 2 Find the net change in entropy when 1.0 kg of water at 100°C are mixed with 1.0 kg of water at 10°C.

first find the final temperature :

$$Q_{lost} = Q_{gained}$$
$$mc\Delta T = mc\Delta T$$
$$(1.0 \text{ kg})(1.0 \text{ kcal/kg-C°})(100°\text{C} - T) = (1.0 \text{ kg})(1.0 \text{ kcal/kg-C°})(T - 10 °\text{C})$$
$$T = 55°\text{C}$$

heat lost = heat gained = $mc\Delta T$ = $(1.0 \text{ kg})(4190 \text{ J/kg-C°})(45 \text{ C°}) = 1.9 \times 10^5$ J

$\Delta S = \Delta S_1 + \Delta S_2$

$\Delta S = [\ (1.9 \times 10^5 \text{ J}) / 305.5 \text{ K}] + [\ (-1.9 \times 10^5 \text{ J}) / 350.5 \text{ K} = 80$ J/K

Example 3 Two 1200-kg cars approach each other at 20 m/s. The cars collide head-on. What is the change in entropoy as a result of the collision on a day when the temperature is 20°C ?

since the cars are moving in opposite directions and the car's have equal mass and speeds, the net momentum after is zero ; therefore all of the kinetic energy goes into the form of heat :

$Q = K_1 + K_2 = (1/2)mv^2 + (1/2)mv^2 = mv^2 = (1200 \text{ kg})(20 \text{ m/s})^2 = 4.8 \times 10^5$ J

$\Delta S = Q / T$; $\Delta S = (4.8 \times 10^5 \text{ J}) / (293 \text{ K}) = 1.6 \times 10^3$ J/K

12.4 Heat Engines and Heat Pumps

Example 1 A heat engine has an efficiency of 30%. It performs work at the rate of 3.5 kW.
A. What is the rate of heat input ?
B. What is the rate of heat discharge ?

given : $\varepsilon = 0.30$; $W = 3.5$ kW

(a) $\varepsilon = W / Q_h$
$0.30 = (3.5 \text{ kW}) / Q_h$; $Q_h = 11.7$ kW
(b) $Q_h = W + Q_c$
$11.7 \text{ kW} = 3.5 \text{ kW} + Q_c = 8.2 \text{ kW}$

Example 2 A heat engine does 4.0×10^3 J of work and exhausts 2.0 kcal of heat. What is the efficiency of the heat engine ?

given : $W = 4.0 \times 10^3$ J ; $Q_c = 2.0 \text{ kcal} = 8.4 \times 10^3$ J

$\varepsilon = W / Q_h$; $Q_h = W + Q_c = 4.0 \times 10^3 \text{ J} + 8.4 \times 10^3 \text{ J} = 12.4 \times 10^3 \text{ J}$
$\varepsilon = (4.0 \times 10^3 \text{ J}) / (12.4 \times 10^3 \text{ J}) = 0.32$ or 32%

Example 3 A refrigerator removes heat from the freezing compartment of a freezer at -25°C and ejects it into a garage whose temperature of 30°C. What is the coefficient of performance ?

given : $T_c = -25°\text{C} = 248$ K ; $T_h = 30°\text{C} = 303$ K

$\text{cop} = T_c / (T_h - T_c)$
$\text{cop} = (248 \text{ K}) / [(303 \text{ K}) - (248 \text{ K})] = 4.5$

12.5 The Carnot Cycle and Ideal Heat Engines

Example 1 A Carnot engine exhausts heat at a temperature of 80°C when the efficiency is 25%. What is the temperature combustion takes place ?

given : $\varepsilon = 0.25$; $T_c = 80°C = 353\ K$

$\varepsilon = 1 - (T_c / T_h)$

$0.25 = 1 - [\ (353) / T_h)\ ]$

$T_h = 471\ K = 198\ °C$

Example 2 A Carnot engine operates under the temperatures of 200°C and 400°C. What is the Carnot efficiency ?

given : $T_c = 200\ °C = 473\ K$; $T_h = 400°C = 673\ K$

$\varepsilon = 1 - (T_c / T_h) = 1 - (473\ K / 673K) = 30\%$

Solutions to paired problems and other selected problems.

2. (d)

10. (b)

16. given : $p = 1.28 \times 10^4$ Pa ; $V_o = 0.25\ m^3$; $V = 0.30\ m^3$; $Q = 200$ cal $= 8.38 \times 10^2$ J

    (a) $W = p\,\Delta V = (1.28 \times 10^4\ Pa)(0.05\ m^3) = 6.40 \times 10^2\ J$
    (b) $Q = \Delta U + W$
    $8.38 \times 10^2 = \Delta U + 6.40 \times 10^2\ J$; $\Delta U = 198\ J$

24. Work done equals the area of the graph :
    $W = (1/2)\ ab + p\Delta V$
    $W = (1/2)(3.0 \times 10^5\ Pa)(0.50\ m^3) + (2.0 \times 10^5\ Pa)(0.50\ m^3) = 1.75 \times 10^5\ J$

30. Since the change in heat for the cold and hot water are the same, the numerators are the same. Yet the temperatures are not the same. Since the hot water is at a higher temperature, the entropy is numerically smaller giving the entire system a positive change in entropy.

36. $Q_1 = mc\Delta T$; $Q_1 = (0.25\text{ kg})(1.0\text{ kcal/kg-C°})(10\text{ C°}) = 2.5\text{ kcal}$

$Q_2 = mL$; $Q_2 = (0.25\text{ kg})(80\text{ kcal / kg}) = 20\text{ kcal}$

$Q_3 = mc\Delta T$; $Q_3 = (0.25\text{ kg})(0.50\text{ kcal/kg-C°})(6.0\text{ C°}) = 7.5\text{ kcal}$

$\Delta S = \Delta S_1 + \Delta S_2 + \Delta S_3$

$\Delta S = (-2.5\text{ kcal / 278 K}) + (-20\text{ kcal / 273 K}) + (-7.5\text{ kcal / 270 K}) = -9.2 \times 10^{-2}\text{ kcal}$

$\Delta S = -4.6 \times 10^2\text{ J/K}$

42. (a) $\Delta S = Q / T$; $273\text{ J/K} = Q / (100\text{ K})$; $Q = 2.73 \times 10^4\text{ J}$
(b) isoentropic since the entropy remains constant at 200 J/K

50. No ; this is a natural process based on the density and buoyancy of air at different temperatures. Since it is a natural process, it results in increased entropy and the second law is not violated.

54. (a) $\varepsilon = W / Q_{in}$; $0.30 = W / 750\text{ J}$; $W = 225\text{ J}$
(b) since useful work is 30% , waste heat must be 70%
0.70 (750 J) = 525 J

58. (a) $cop = Q_{out} / W_{in}$; $cop = (250\text{ kcal}) / (150\text{ kcal}) = 1.67$
(b) $W = Q_{hot} - Q_{cold}$; $W = (400\text{ kcal}) - (250\text{ kcal}) = 150\text{ kcal or } 6.29 \times 10^5\text{ J}$

72. $\varepsilon = 1 - (T_c / T_h)$; $\varepsilon = 1 - (393\text{ K / 593 K})$; $\varepsilon = 34\ \%$

$\varepsilon = W / Q_h$; $0.34 = W / (2.7 \times 10^4\text{ J})$; $W = 9.3 \times 10^3\text{ J}$

76. $\varepsilon = 1 - (T_c / T_h)$

$\varepsilon_1 = 1 - (373\text{ K / 573 K}) = 0.35 = 35\ \%$

$\varepsilon_2 = 1 - (100\text{ K / 300 K}) = 0.67 = 67\ \%$

Sample Quiz

Multiple Choice. Choose the correct answer.

__ 1. The first law of thermodynamics is another statement for
A. law of conservation of mass
B. law of conservation of energy
C. law of conservation of momentum
D. law of conservation of angular momentum

__ 2. A refrigerator operates with a compartment temperature of 35°F and rejects heat into a room. At what room temperature is the coefficient of performance the greatest ?
A. 50°F B. 65°F C. 76°F D. 90°F

__ 3. What is the Carnot efficiency of a heat engine operating between the temperatures of 50°C and 200°C ?
A. 25% B. 32% C. 75% D. 68%

__ 4. In an adiabatic process the change in _____ is zero
A. temperature B. internal energy C. work D. heat

__ 5. When 0.50 kg of ice melts, the change in entropy is
A. 0.40 kcal/K B. 0.80 kcal/K C. 0.15 kcal/K D. 0.11 kcal/K

__ 6. The work done on a system in a isothermal process is -400 J. What is the change in internal energy ?
A. -400 J B. 0 C. 400 J D. 800 J

__ 7. In an isothermal process, there is no change in the
A. pressure B. volume C. temperature D. heat

Problems

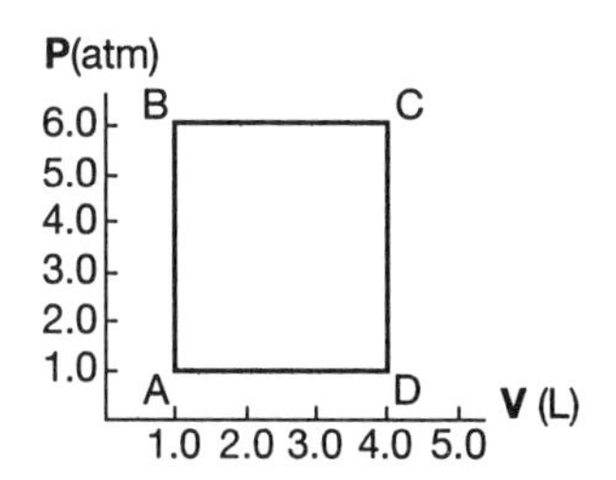

Fig.12.3

8. In the process cycle in Fig. 12.3, two moles of an ideal gas undergoes a process ABCDA.
   A. Find the work done for a complete cycle.
   B. Find the change in internal energy for a complete cycle.
   C. Find the change in heat for a complete cycle.
   D Find the temperature at point C.

9. One kilogram of water is cooled from room temperature to 10°C. What is the change in entropy ?

10. A Carnot engine operates between the temperature of 300°C and 500°C and discharges heat at the rate of 5.0 kW.
    A. Find the Carnot efficiency of the engine.
    B. Find the rate of work output.
    C. Find the rate of heat input.

CHAPTER 13 Vibrations and Waves

Chapter Objectives

Upon completion of the unit on vibrations and waves, students should be able to :

1. describe the function of a wave.
2. state and apply Hooke's law.
3. describe periodic motion and simple harmonic motion.
4. find the period, frequency, wavelength, and speed of a wave.
5. find the potential, kinetic, and total energies of a wave undergoing simple harmonic motion.
6. write equations for the position, velocity, and acceleration as functions of time for simple harmonic motion, and write expressions for the maximum velocity and acceleration.
7. write expressions for the period, angular frequency, and frequency for bodies undergoing SHM in terms of k and the mass of the oscillating body.
8. explain the concept of sinusoidal motion.
9. explain damping.
10. sketch graphs of the position, velocity, and accelerations as functions of time for SHM.
11. sketch graphs of the potential, kinetic, and total energy as functions of the position and time for SHM.
12. write an expression for the period of a simple pendulum, noting the variables which affect the pendulum's period.
13. distinguish between a transverse and a longitudinal wave.
14. explain the principle of superposition in order to explain the phenomena of interference and the various types of interference.
15. explain the properties of reflection, refraction, and diffraction for wave motion.
16. explain the properties of standing waves.
17. define fundamental frequency, and define harmonics in terms of the fundamental frequency.
18. write and expression for the wave speed in a stretched string.

Chapter Summary

*An oscillation or a vibration requires a restoring force. Oscillatory motion under the influence of a force is describes by Hooke's law ($F = -kx$) is called **simple harmonic motion (SHM)**. SHM is described by the parameters of displacement (x), amplitude (A), period (T), and frequency (f). The displacement is described by a sinusoidal (sine and/or cosine) function of time. The unit of frequency is the hertz (Hz).

*The total mechanical energy of an object in SHM is directly proportional to the square of its amplitude (A). For example, for a mass m oscillating on a spring, $E = (1/2)kA^2$, where k is the spring constant. The maximum speed of the mass is directly proportional to the amplitude, $v_{max} = A\,[k/m]^{1/2}$.

*The **equation of motion** of an object in SHM has the same general form $y = A \sin(\omega t + \delta)$, where ω is the angular frequency ($\omega = 2\pi f = 2\pi/T$) and δ is the phase constant. The angular frequency is determined by the material parameters of the system, e.g., $\omega = [k/m]^{1/2}$ for a mass on a spring. The amplitude and phase constant depend in the initial conditions. Being a sine function, a cosine function, or a combination of both, the equation of motion is sinusoidal.

*With frictional losses and without a driving force, the amplitude and energy of an oscillator decrease with time, and its motion is said to be damped.

***Wave motion** is the propagation of a disturbance or a process of energy transfer. A wave is characterized by its amplitude, wavelength, and frequency. The **wave speed** is given by $v = \lambda f$.

*Waves are divided into two categories or types based on the directions of particle oscillations relative to the wave velocity. For a **transverse wave**, the particle motion is perpendicular to the direction of the wave velocity. A transverse wave is sometimes called a shear wave. For a **longitudinal wave**, the particle motion is parallel to the direction of the wave velocity. A longitudinal wave is sometimes called a compressional wave.

***Interference** occurs when waves meet or overlap. By the **principle of superposition**, at any time, the combined waveform of two or more interfering waves is given by the sum of the displacement of the individual waves at each point in the medium. If the resulting wavefront is greater in amplitude than any of the interfering waves, we have what is called **constructive interference**. If the amplitude of the combined wave form is smaller than any of the individual waves, we have what is called **destructive interference**. Total destructive interference occurs when interfering waves of equal frequency and amplitude are completely 180° out of phase and total constructive interference occurs when two interfering waves of equal frequency and amplitude are exactly in phase.

*__Reflection__ occurs when a wave strikes an object or comes to the boundary of another medium and part of the disturbance is diverted back into the original medium.

*__Refraction__ occurs when a wave crosses a boundary into another medium and the transmitted wave moves in a different direction.

*__Diffraction__ is the bending of a wave around an object or a corner.

*Interfering waves traveling in opposite directions in a string or rope produce a **standing wave**. The string does not undergo a displacement at certain points called **nodes**, and the standing wave has its maximum amplitude at points called **antinodes**. The oscillating pattern of a standing wave stands in place with fixed nodes and antinodes.

*The frequencies at which large -- amplitude standing waves are produced are called **natural** or **resonant frequencies** (or sometime characteristic frequencies). The lowest resonant frequency f_1 is called the **fundamental frequency** or the first harmonic. The set of natural frequencies is called a harmonic series, with the next higher frequency f_2 the second harmonic, etc.

Important Terms and Relationships

13.1 Simple Harmonic Motion

Hooke's Law : $F = -kx$
oscillation (vibration)
simple harmonic motion (SHM)
displacement
amplitude
period : $T = 1/f$
frequency
hertz (Hz)
total energy of a spring and mass in SHM ; $E = (1/2)kA^2 = (1/2)mv^2 + (1/2)kx^2$
velocity of oscillating mass : $v = \pm[(k/m)(A^2 - x^2)]^{1/2}$
maximum speed of oscillating mass : $v = (k/m)^{1/2}(A)$
period of mass oscillating on a spring : $T = 2\pi(m/k)^{1/2}$

frequency of mass oscillating on a spring : $f = (1 / 2\pi) (k / m)^{1/2}$
period of a simple pendulum : $T = (2\pi) (L/g)^{1/2}$

13.2 Equations of Motion

equations of motion

displacement of a mass in SHM : $y = A \sin (\omega t + \delta)$
velocity of a mass in SHM ($\delta = 0$) : $v = A\omega \cos (\omega t + \delta)$
acceleration of mass in SHM ($\delta = 0$) : $a = -A\omega^2 \sin (\omega t + \delta)$

angular frequency : $\omega = 2\pi f = (k/m)^{1/2}$
phase constant
damped harmioc motion

13.3 Wave Motion

wave motion
wave
wave pulse
periodic wave
wavelength (λ)
wave speed : $v = \lambda f$
transverse wave
longitudinal wave

13.4 Wave Phenomena

principle of superposition
constructive interference
destructive interference
total constructive interference
total destructive interference
reflection
refraction
diffraction

13.5 Standing Waves and Resonance

standing wave
node
antinode
natural (resonant) frequencies : $f_n = nv / 2L = (n / 2L)(F / \mu)^{1/2}$ (for n = 1, 2, 3, ...)
fundamental frequency
harmonic series
resonance

Additional Solved Problems

13.1 Simple Harmonic Motion

Example 1 A 0.50-kg mass moves at 1.5 m/s along a horizontal, frictionless surface and strikes a spring whose force constant is 20 N/m.
A. What is the total energy of the system ?
B. What is the amplitude ?
C. At what location are the values for the potential and kinetic energies the same ?

given : $m = 0.50$ kg ; $v_{max} = 1.5$ m/s ; $k = 20$ N/m

A. $E = K_{max} = (1/2)\, mv_{max}^2 = (1/2)(0.50 \text{ kg})(1.5 \text{ m/s})^2 = 0.56$ J

B. $U_{max} = (1/2)(20 \text{ N/m})(A^2)$; $0.56 \text{ J} = (1/2)(20 \text{ N/m})\, A^2$; $A = 0.24$ m

C. $E = K + U$; $0.56 \text{ J} = K + K$; $K = U = 0.28$ J ;
$U = (1/2)\, kx^2$; $0.28 \text{ J} = (1/2)(20 \text{ N/m})\, x^2$; $x = 0.17$ m

Example 2 A 0.050-kg is suspended from a spring with a force constant is 20 N/m.
A. How far will the mass stretch the spring ?
B. The mass is now set into motion. What is the equlibrium position as the mass moves up and down ?

given : $m = 0.050$ kg ; $k = 20$ N/m

A. $F = kx$; $F = mg$
$kx = mg$
$(20 \text{ N/m})\, x = (0.050 \text{ kg})(9.80 \text{ m/s}^2)$; $x = 2.5 \times 10^{-2}$ m or 2.5 cm

B. the equilibrium position for a vertical spring occurs when the net force is zero ; from part (A) the mass oscillates around the 2.5 cm position.

13.2 Equations of Motion

Example 1 A mass of 10 kg oscillates when attached to a spring moves on a frictionless horizontal surface. It is observed the period of the motion is 0.20 s and the amplitude of the mass is 30 cm.

A. What is the force constant of the spring ?
B. What is the total energy of the system ?

given : m = 0.10 kg ; T = 0.20 s ; A = 0.30 m

A. $T = 2\pi\,(m / k)^{1/2}$

$0.20 = 2\pi\,(0.10\text{ kg} / k)^{1/2}$

$0.040 = 4\,\pi^2\,(0.10 / k)$; $k = 99$ N/m

B. $E = U_{max} = (1/2)kA^2$; $E = (1/2)(99\text{ N/m})(0.30\text{ m})^2 = 4.5$ J

Example 2 A pendulum with a 0.50 kg bob has a length of 50 cm.

A. What is the frequency of the pendulum ?
B. What is the period of the pendulum
C. What is the period if the 0.50-kg mass is replaced with a 1.0-kg mass ?

given : m = 0.50 kg ; L = 0.50 m

find (B) first : $T = 2\pi\,(L / g)^{1/2}$; $T = 2\pi\,[\,(0.50\text{ m}) / (9.80\text{ m/s}^2)\,]^{1/2} = 1.4$ s

A. $f = 1 / T$; $f = 1 / (1.4\text{ s}) = 0.70$ Hz
C. notice the mass is independent of the period ; the period remains 1.4 s

Example 3 A 0.50-kg mass undergoes SHM according to the equation :

$$x = (10\text{ cm})\sin(\pi t)$$

A. What is the amplitude ?
B. What is the frequency ?
C. What is the period ?
D. What is the maximum speed of the mass ?

A. A = 10 cm

B. $\omega = \pi$ rad/s ; $\omega = 2\pi f$; $(\pi\text{ rad/s}) = (2\pi) f$; $f = 0.50$ Hz

C. $T = 1 / f$; $T = 1 / 0.50 = 2.0$ s

D. $v_{max} = A\,\omega$; $v_{max} = (10\text{ cm})(3.14\text{ rad/s}) = 31.4$ cm/s

13.3 Wave Motion

Example 1 The frequency of a longitudinal wave is 340 Hz and the speed of the wave is 340 m/s.
A. What is the period of the wave ?
B. What is the wavelength ?

given : $f = 340$ Hz ; $v = 340$ m/s

A. $T = 1 / f$; $T = 1 / (340 \text{ Hz}) = 2.9 \times 10^{-3}$ s

B. $v = \lambda f$; $(340 \text{ m/s}) = \lambda (340 \text{ Hz})$; $\lambda = 1.00$ m

Example 2 Violet light has a wavelength of 4.0×10^{-7} m and a speed of 3.0×10^{8} m/s. What is the frequency of the light ?

given : $v = 3.0 \times 10^{8}$ m/s ; $\lambda = 4.0 \times 10^{-7}$ m

$v = \lambda f$; $(3.0 \times 10^{8} \text{ m/s}) = (4.0 \times 10^{-7} \text{ m}) f$; $f = 7.5 \times 10^{14}$ Hz

13.5 Standing Waves and Resonance

Example 1 A 60-Hz oscillator vibrates a string 1.0 m long with three loops. Find the speed of the waves in the string.

given : $f = 60$ Hz ; $L = 1.0$ m ; $n = 3.0$

$f_n = nv / 2L$; $60 \text{ Hz} = (3) v / 2 (1.0 \text{ m})$; $v = 40$ m/s

Example 2 A 20-m length string mas a mass of 1.0 g. A 2.0-m segment of the string is fixed at both ends. When a tension of 20 N is applied to the string, four loops are produced. What is the frequency of the wave ?

given : $L = 20$ m ; $m = 1.0$ g ; $l_o = 2.0$ m ; $T = 20$ N

$\mu = m / L = (1.0 \times 10^{-3} \text{ kg}) / (2.0 \text{ m}) = 5.0 \times 10^{-4}$ kg / m

$v = [T / \mu]^{1/2}$; $v = [(20 \text{ N}) / (5.0 \times 10^{-4} \text{ kg/m})]^{1/2} = 2.0 \times 10^{2}$ m/s

$f_n = nv / 2L$; $f_4 = (4)(2.0 \times 10^{2} \text{ m/s}) / 2(2.0 \text{ m}) = 2.0 \times 10^{2}$ Hz

Solutions to paired problems and other selected problems.

2. the total energy $E = (1/2)kA^2$; (c)

8. total energy : $E = U + K$; at equilibrium $U = 0$ and v is the maximum speed

$E = 0 + (1/2)\ mv^2$ where $v_{max} = A\omega$

$E = (1/2)\ m\ (A\ \omega)^2 = (1/2)\ m\omega^2A^2$

14. (a) $U = (1/2)\ kx^2$; $U = (1/2)(75\ N/m)(0.15\ m)^2 = 0.84\ J$ as measured from the equilibrium ; gravitational potential energy is also present but it depends upon where the zero height level is chosen

(b) yes, since the mass is oscillating in a vertical plane when gravitational potential energy is taken into account. Since the author says to ignore gravitational potential energy the answer to the questions is **no**.

21. $\omega = (k / m)^{1/2}$; the answer if (d) ;

(a) is wrong because the angular frequency is inversely proortional to the square root of the mass.

(b) the angular frequency and the period are inversely proportional.

(c) no relationship between the phase constant and the angular frequency.

28. $y = (7.0\ cm)\ \cos\ (10\ t)$

first the angular frequency is 10 rad/s ; $\omega = 2\pi f$; $10\ rad/s = 2\pi\ (f)$; $f = 1.6\ Hz$

(a) $T = 1 / f$; $T = 1 / (1.6\ Hz) = 0.63\ s$

(b) the phase constant is 90°

(c) $T = 2\pi\ [m/k]^{1/2}$; $(0.63\ s)^2 = 4\pi^2\ [(0.010\ kg) / k)]^{1/2}$; $k = 0.99\ N/m$

$E = U_{max} = (1/2)\ kA^2 = (1/2)(0.99\ N/m)(0.07)^2 = 2.4 \times 10^{-3}\ J$

32. from the graph the period is 8.0 s ; $f = 1 / T$; $f = 1 / (8.0\ s) = 0.13\ Hz$

$\omega = 2\pi f = 2\pi\ (0.13\ Hz) = 0.79\ rad/s$ or $(\pi/4)\ rad/s$;

$y = (5.0\ cm)\ \cos\ (\pi\ t /4)$

$y = (5.0\ cm)\ \cos\ (2\pi\ t / 8)$

$y = (5.0\ cm)\ \sin\ [\ (\pi t / 4) + (\pi/2)]$

43. (a) transverse and longitudinal waves will propagate in solids.

(b) longitudinal waves will propagate in fluids.

(c) transverse waves will propagate in stretched strings.

46. $v = \lambda f$; $v = 3{,}000{,}000 \text{ km / s} = 3.0 \times 10^8 \text{ m/s}$
$(3.0 \times 10^8 \text{ m/s}) = \lambda\,(10^{14} \text{ Hz})$; $\lambda = 3.0 \times 10^{-6} \text{ m}$

52. $v = [\, Y / \rho \,]^{1/2}$
(a) $v = [\,(7.0 \times 10^{10} \text{ N/m}^2) / (2.7 \times 10^3 \text{ kg/m}^3)\,]^{1/2}$; $v = 5.1 \times 10^3 \text{ m/s}$
$v = \lambda f$; $(5.1 \times 10^3 \text{ m/s}) = \lambda\,(40 \text{ Hz})$; $\lambda = 1.3 \times 10^2 \text{ m}$
(b) $v = [\,(11 \times 10^{10} \text{ N/m}^2) / (8.9 \times 10^3 \text{ kg/m}^3)\,]^{1/2}$; $v = 3.5 \times 10^3 \text{ m/s}$
$(3.5 \times 10^3 \text{ m/s}) = \lambda\,(40 \text{ Hz})$; $\lambda = 88 \text{ m}$

60. (c)

66. (a) $v = [\, T / \mu]^{1/2}$; $v = [\,(9.0 \text{ N}) / (0.125 \text{ kg/m})\,]^{1/2}$; $v = 8.5 \text{ m/s}$
(b) $f_n = nv / 2L = (n / 2L)(F / \mu)^{1/2}$; $f_n = (1)(8.5 \text{ m/s}) / 2(10 \text{ m})$; $f_n = 0.43\,(n) \text{ Hz}$

72. 1.5 loop means that (3/4) wavelength is present :
$(3/4)\,\lambda = 3.0 \text{ m}$; $\lambda = 4.0 \text{ m}$

83. $y = (0.20 \text{ cm}) \sin (1.8\,\pi t)$; $v = (0.20 \text{ cm})(1.8\,\pi) \cos (1.8\,\pi t)$
$v = (0.20 \text{ cm})(1.8\,\pi) \cos [(1.8)\,\pi\,(10 \text{ s})]$
$v = 1.1 \text{ cm/s}$

Chapter 13

Sample Quiz

Multiple Choice. Choose the correct answer.

__ 1. The length of simple pendulum is doubled. By what factor is the period of the pendulum changed ?

A. $1 / (2)^{1/2}$ B. 1 / 2 C. $(2)^{1/2}$ D. 2

__ 2. The period of a pendulum on Earth is 1.0 s. The same pendulum is now taken to a hypothetical planet whose acceleration due to gravity is 4.0 m/s^2. What is the period of the pendulum on the hypothetical planet ?

A. 0.40 s B. 0.63 s C. 1.6 s D. 2.5 s

__ 3. When a mass m oscillates on a spring, its period is observed to be T. A mass 2 m is now placed on the spring and it again obeys SHM. What is its period ?

A. T / 2 B. T / 1.4 C. (1.4) T D. 2 T

__ 4. The speed of a wave is observed to be 20 m/s and its wavelength is observed to be 4.0 m. What is the period of the wave ?

A. 4.0 s B. 20 s C. 0.20 s D. 5.0 s

__ 5. A stiing is observed to be vibrating as an entire unit with nodes on the fixed ends. If the distance between the fixed ends is 2.0 m, what is the wavelength ?

A. 0.50 m B. 1.0 m C. 2.0 m D. 4.0 m

__ 6. The tension in a string is quadrupled. What happens to the frequency assuming the wavelength stays the same ?

A. 4 times greater B. doubled C. halved D. quartered

__ 7. The period of a wave is doubled. What happens to the frequency ?

A. 4 times greater B. doubled C. halved D. quartered

Problems

8. The position of a 0.20-kg mass in SHM is given by the the equation :

$$x = (2.0\ \text{cm}) \cos(\pi\ t/2)$$

A. What is the amplitude ?
B. What is the maximum speed of the mass ?
C. What is the spring constant ?
D. What is the period of an oscillation ?

9. Two physics students observe water waves on a lake. One student observes that 10 waves pass a given point in 30 s. A second student observes it takes 30 s for the wave to travel a distance of 150 m. What is the wavelength of the water waves ?

10. A string has a mass of 2.0 g and has a length of 2.0 m. When subjected to a tension of 20.0 N and set into motion it is observed that four loops exists between the fixed ends.
A. What is the speed of the wave ?
B. What is the frequency of the wave ?

CHAPTER 14 Sound

Chapter Objectives

Upon completion of the unit on sound, students should be able to :

1. define sound waves.
2. classify the three regions of the sound spectrum into three regions according to frequency.
3. list some practical applications of ultrasound.
4. calculate the speed of sound for various materials.
5. apply the expression for the speed of sound in air.
6. calculate the intensity and measure the intensity level of sound.
7. explain and apply the properties of reflection, refraction, and diffraction of sound.
8. using the path difference of waves, determine the type interference present, or find locations of constructive or desructive interference.
9. calculate the beat frequency.
10. apply the Doppler effect for moving sound sources and / or moving observers.
11. explain why a sonic boom occurs.
12. calculate the speeds of objects in the form of Mach numbers.
13. explain the sensory effects of loudness, pitch, and quality.
14. explain the concept of resonance.
15. apply the expressions for stringed instruments and for open and closed tubes.

Chapter Summary

*The high- and low- density pressure regions (crests and troughs) of a longitudinal sound wave are called condensations and rarefactions, respectively.

*The sound spectrum has three regions: the infrasonic region ($f < 20$ Hz), the audible region ($20 < f < 20$ kHz), and the ultrasonic region ($f > 20$ kHz)

*The **speed of sound** in a medium depends on the elasticity or the intermolecular interactions of the medium and the mass or the density of its particles. The speed of sound in air is temperature dependent and increases by about 0.6 m/s per degree Celsius over the normal temperature range. In general, the speed of sound in air is on the order of 1/3 km/s (or 1.5 mi/s).

***Sound intensity** is the acoustical energy transported per units time per area : intensity I = energy/time/area = power/area (W/m^2). For a point source, the intensity is inversely proportional to the square of the distance from the source (an inverse square relationship).

*Sound intensity is perceived as loudness. The threshold of hearing at 1000 Hz is $I_o = 10^{-12}$ W/m^2. The threshold of pain is I = 1.0 W/m^2. The sound level intensity (β) is expressed on a relative logarithmic scale with the unit of **decibel** (dB). Sound intensity levels are referenced to I_o and $\beta = 10\log(I/I_o)$ dB (base-10 log). The dB range between the thresholds of hearing and pain is then between 0 dB and 120 dB.

*Sound waves may be reflected, refracted, and diffracted : that is, bounce off objects or surfaces , have a change in direction due to a medium or density change, and be bent around corners or objects, respectively.

*Sound waves interfere, and spatially, this depends on the path difference of the waves. If the path difference of two waves with the same frequency is zero or an integral (whole) number of wavelengths ($n\lambda$, with n = 0,1,2,3,...), constructive interference occurs. If the path difference is an odd number of half-wavelengths ($m\lambda$ = 1,3,5, ...), destructive interference occurs.

*When two tones of nearly the same frequency ($f_1 \approx f_2$) interfere, pulsations in loudness known as **beats** result. For two sinusoidal waves of equal amplitude, the beat frequency is given by the difference in the wave frequencies, $f = | f_1 - f_2 |$.

*Variations in the observed frequency of a sound source because of motions of the source and/or the observer result from what is known as the **Doppler effect**. The Doppler effect also occurs for electromagnetic waves. A Doppler “red shift” from light from galaxies is evidence for the theory of an expanding universe, and traffic radar is based on the Doppler effect.

*Objects traveling at supersonic speeds produce large pressure ridges or shock waves. This gives risk to the **sonic boom** heard with the passing of a supersonic jet aircraft. The ratio of the speed of sound v and the speed of the source v_s is called the Mach number ($M = v_s / v$) and $M > 1$ for supersonic speeds.

*The sensory effects of loudness, pitch, and quality are related to the physical wave properties of intensity, frequency, and waveform (harmonics), respectively.

*Stringed musical instruments produce notes by setting up standing waves in strings with different fundamental frequencies. Standing waves (longitudinal) may be set up in air columns, and this is the basis for producing different notes in organ pipes and wind instruments.

Important Terms and Relationships

14.1 Sound Waves

sound waves
sound frequency spectrum
audiable region
infrasonic region
ultrasonic region

14.2 Speed of Sound

speed of sound in air (m/s) : $v = 331 + (0.6)T_C$

speed of sound in different media : $v = [\, Y / \rho]^{1/2}$

14.3 Sound Intensity

intensity

intensity of point source : $I = P / 4\pi R^2$

decibel (dB)

intensity level (in dB) : $\beta = 10 \log (I / I_o)$; $I_o = 10^{-12}$ W/m^2

sound intensity level

14.4 Sound Phenomena

constructive interference
phase difference
path difference

condition for constructive interfèrence: $PD = n\lambda$, when n = 0, 1, 2, 3, ...
destructive interference
condition for destructive interference $PD = m(\lambda/2)$, when m =1,3, 5, ...
beats
beat frequency : $f_b = | f_1 - f_2 |$
Doppler effect : $f_o = [v / (v \pm v_s)] f_s$ or $f_o = [(v \pm v_o) / v] f_s$
Mach number (M) : $M = 1 / \sin\theta = v_s / v$

14.5 Sound Characteristics

loudness
pitch
quality
natural frequencies (open pipe) : $f_n = nv / 2L$ for n = 1, 2, 3, ...
natural frequencies (closed pipe) : $f_m = mv / 4L$ for m = 1, 3, 5, ...

Additional Solved Problems

14.1 Sound Waves

Example An ultrasonic wave has a frequency of 30 kHz.
A. What is the period of the wave ?
B. If the speed of the wave is 500 m/s, what is the wavelength ?

A. $T = 1 / f$; $T = 1 / (30 \times 10^3 \text{ Hz})$ (since k or kilo means 1000)
$T = 3.3 \times 10^{-5}$ s
B. $v = \lambda f$; $500 \text{ m/s} = \lambda (30 \times 10^3 \text{ Hz})$; $\lambda = 1.7 \times 10^{-2}$ m

14.2 Speed of Sound

Example 1
A. Determine the speed of sound in air when the temperature is 30°C .
B. What is the wavelength of a sound wave whose frequency is 440 Hz ?

given : $f = 30 \times 10^3$ Hz ; v = 500 m/s . T_C = 30°C
A. $v = [331 + (0.6) T_C]$ m/s
$v = [331 + (0.6)(30)] = 349$ m/s

B. $v = \lambda f$; 349 m/s = λ (440 Hz) ; λ = 0.79 m

Example 2 Find the speed of sound in an aluminum rod ?

given : $Y_{Al} = 7.0 \times 10^{10}$ N/m^2 ; $\rho_{Al} = 2.7 \times 10^3$ kg/m^3

$v = [Y / \rho]^{1/2}$; $v = [(7.0 \times 10^{10} \text{ N/m}^2) / (2.7 \times 10^3 \text{ kg/m}^3)]^{1/2} = 5.1 \times 10^3$ m/s

Example 3 A sound is produced in a valley and an echo is heard 5.0 s later. If the air temperature is 20°C, what is the distance from where the sound is produced to the point where it is reflected ?

given : T_C = 20°C ; t = 5.0 s

$v = [331 + (0.6)T_C]$ m/s

$v = [331 + (0.6)(20)]$ m/s = 343 m/s

x = vt ; since the echo is heard 5.0 s later, it takes 2.5 s from the mass to get to the point where it is reflected and 2.5 s for it to return.

x = (343 m/s)(2.5 s) = 8.6×10^2 m

14.3 Sound Intensity

Example 1 The intensity of a sound produced by a speaker is 5.0×10^{-2} W/m^2 at a distance of 2.0 m from a source. What is the intensity level 6.0 m from the source ?

given : I = 5.0×10^{-2} W/m^2 ; d_1 = 2.0 m ; d_2 = 6.0 m ;

$I_1 / I_2 = r_2^2 / r_1^2$

$(5.0 \times 10^{-2} \text{ W/m}^2) / I_2 = [(6.0 \text{ m}) / (2.0 \text{ m})]^2$; $I_2 = 5.5 \times 10^{-3}$ W/m^2

Example 2 Find the intensity level for a sound with an intensity 3.0×10^{-1} W/m^2.

given : $I_o = 1.0 \times 10^{-12}$ W/m^2 ; I = 3.0×10^{-1} W/m^2

$\beta = 10 \log (I / I_o)$; $\beta = 10 \log [(3.0 \times 10^{-1} \text{ W/m}^2) / (10^{-12} \text{ W/m}^2)]$

$\beta = 10 \log (3.0 \times 10^{11})$; β = 115 dB

Example 3 The intensity level of two tones are 70 dB and 75 dB. What is the intensity level of the combination if they are produced at the same time ?

given : $\beta_1 = 70$ dB ; $\beta_2 = 75$ dB ; $I_o = 10^{-12}$ W/m^2

$\beta = 10 \log (I / I_o)$;

$70 = 10 \log (I / I_o)$; $7 = \log (I / 10^{-12})$; $1 \times 10^7 = I / (10^{-12}$

$I_1 = 1.0 \times 10^{-5}$ W/m^2

$7.5 = 10 \log (I / I_o)$; $3.2 \times 10^7 = I_2 / 10^{-12}$ W/m^2 ; $I_2 = 3.2 \times 10^{-5}$ W/m^2

$I = I_1 + I_2 = (1.0 \times 10^{-5}$ W/m$^2) + (3.2 \times 10^{-5}$ W/m$^2) = 4.2 \times 10^{-5}$ W/m^2

$\beta = 10 \log [(4.2 \times 10^{-5}$ W/m$^2) / (10^{-12}$ W/m$^2)]$; $\beta = 76$ dB

14.4 Sound Phenomena

Example 1 A person stands between two speakers. Each speaker produces a tone with a frequency of 200 Hz on a day when the speed of sound is 330 m/s. The person is 1.65 m from one speaker and 4.95 m from the other speaker. What type of interference is located at this point ?

given : $v_s = 330$ m/s ; $f = 200$ Hz ; $d_1 = 1.65$ m ; $d_2 = 4.95$ m

$v = \lambda f$; 330 m/s $= \lambda$ (200 Hz) ; $\lambda = 1.65$ m

now take the difference in the distances between the speakers
4.95 m - 1.65 m = 3.30 m

3.30 m = 2 (λ) ; therefore there is constructive interference.

Example 2 Two musical notes are played by instruments, one at 350 Hz and the other at 355 Hz. What is the beat frequency ?

given : $f_1 = 350$ Hz ; $f_2 = 355$ Hz

$f_b = | f_1 - f_2 |$; $f_b = 355$ Hz - 350 Hz = 5 Hz

Example 3 A train approaches and passes a crossing at a speed of 80 km/h. The whistle produces a tone with a frequency of 500 Hz. Assume the speed of sound is 340 m/s.

A. For a stationary bystander, what is the pitch heard when
(1) the train approaches.
(2) after the train passes by the crossing.

B. For a car moving parallel to and in the opposite direction of the train at a speed of 40 km/h, what is the pitch heard when
(1) the train and car are approaching each other ?
(2) when the train passes the car ?

A. given : $v = 340$ m/s ; $v_s = 80$ km/h ; $f = 500$ Hz

$v_s = 80$ km / h = 22 m/s

(1) $f_o = [v / (v - v_s)] f_s$; $f_o = \{ (340 \text{ m/s}) / [(340 \text{ m/s}) - (22 \text{ m/s})] \} (500 \text{ Hz}) = 535 \text{ Hz}$

(2) $f_o = [v / (v + v_s)] f_s$; $f_o = \{ (340 \text{ m/s}) / [(340 \text{ m/s}) + (22 \text{ m/s})] \} (500 \text{ Hz}) = 470 \text{ Hz}$

B. given : $v_o = 40$ km/h ; $v = 340$ m/s ; $f = 500$ Hz

(1) $f_o = [(v + v_o) / (v - v_s)] f_s$;

$f_o = \{ [(340 \text{ m/s}) + (11 \text{ m/s})] / [(340 \text{ m/s}) - (22 \text{ m/s})] \} (500 \text{ Hz}) = 552 \text{ Hz}$

(2) $f_o = [(v - v_o) / (v + v_s)] f_s$;

$f_o = \{ [(340 \text{ m/s}) - (11 \text{ m/s})] / [(340 \text{ m/s}) + (22 \text{ m/s})] \} (500 \text{ Hz}) = 454 \text{ Hz}$

14.5 Sound Characteristics

Example 1 An organ pipe 2.0 m long is open on both ends. What is the frequency of the fundamental and the first two overtones if the speed of sound is 340 m/s ?

given : $L = 2.0$ m ; $v = 340$ m/s

$f_n = nv / 2L$;

fundamental : $f_1 = (1.0)(340 \text{ m/s}) / 2 (2.0 \text{ m}) = 85 \text{ Hz}$

first overtone (second harmonic) = 2(85 Hz) = 1.7×10^2 Hz

second overtone (third harmonic) = 3 (85 Hz) = 2.6×10^2 Hz

Example 2 A closed tube resonates with a fundamental frequency of 440 Hz. If the speed of sound is 340 m/s, what is the length of the pipe ?

given : f = 440 Hz ; v = 340 m/s

$f_m = mv / 2L$; 440 Hz = (1.0)(340 m/s) / 2 L ; L = 0.39 m

Example 3 An open organ pipe 1.00 m long resonates at a temperature of 5°C and again at 25°C. Assume the expansion of the pipe is negligible, what is the frequency of the fundamental for each temperature ?

given : L = 1.00 m ; T_1 = 5°C ; T_2 = 25°C

for 5°C : v = $[331 + (0.6)T_C]$ m/s = [331 + (0.6)(5)] m/s = 334 m/s
$f_n = nv / 2L$; f_n = (1)(334 m/s) / 2(1.00 m) = 167 Hz

for 25°C : v = $[331 + (0.6)T_C]$ m/s = [331 + (0.6)(25)] = 346 m/s
f_n = (1)(346 m/s) / 2 (1.00 m) = 173 Hz

Solutions to paired problems and other problems

5. The vibations for the moving wings must be in the audible range - to be heard. ultrasonic or infrasonic -- vibrations not heard.

18. v = $[331 + (0.6)T_C]$ m/s ; v = [331 + (0.6)(16)] m/s = 341 m/s
x = vt ; x = (341 m/s)(0.40 s) = 1.4×10^2 m

22. v = $[331 + (0.6)T_C]$ m/s ; v = [(331 m/s) + (0.6)(100°C) = 391 m/s
T_C is not a normal enviromental temperature.
For better results use
v = (331 m/s) $[1 + (T_C / 273)]^{1/2}$; v = 331 $[1 + (100 / 273)]^{1/2}$ = 386 m/s

29. Yes, there can be negative instensity level measurements. Since all measurements are relative to the threshold of hearing, the negative value would indicate the intensity would be less than the threshold of hearing.

38. $\beta = 10 \log (I / I_o)$; first find I for both balloon :

$100 \text{ dB} = 10 \log (I / I_o)$; $10^{10} = I / (10^{-12} \text{ W/m}^2)$; $I = 10^{-2} \text{ W/m}^2$

$I_{\text{for each}} = 5.0 \times 10^{-3} \text{ W/m}^2$; $\beta = 10 \log [(5.0 \times 10^{-3}) / (10^{-12})]$; $\beta = 97 \text{ dB}$

44. $\log (I / I_o) = \log 10^x = x$ bels

$\text{dB} = 10 \log (10^x) = 10\,x$;
$x \text{ B} = 10\,x \text{ dB}$; $1 \text{ dB} = (1/10) \text{ dB}$

54. (a) $16.50 \text{ m} - 13.50 = 3.00$,
$PD / \lambda = 3.00 \text{ m} / 1.50 \text{ m} = 2$; since whole number there is constructive interference.

(b) $11.25 \text{ m} - 9.00 \text{ m} = 2.25 \text{ m}$; $PD / \lambda = 2.25 \text{ m} / 1.50 \text{ m} = 1.5$; since the (1/2) wavelength difference there is destructive interference.

58. (a) $v_s = 90 \text{ km/h} = 25 \text{ m/s}$; $v = [331.5 + (0.6)(25)] \text{ m/s} = 346 \text{ m/s}$

$f_o = \{1 / [1 - (v_s/v)] \} f_s$;

$f_o = \{ 1 / [1 - (25 \text{ m/s} / 346 \text{ m/s})] \} (4.0 \times 10^2 \text{ Hz})$

$f_o = 429 \text{ Hz}$

(b) $f_o = \{1 / [1 + (v_s/v)] \} f_s$

$f_o = \{ 1 / [1 + (25 \text{ m/s} / 346 \text{ m/s})] \} (4.0 \times 10^2 \text{ Hz})$

$f_o = 373 \text{ Hz}$

62. $M = 1 / \sin \theta$; $M = 1.0$; $1.0 = 1 / \sin \theta$; $\theta = 90°$

70. the quality of sound depends upon the waveform.

77. $f_m = mv / 4L$;

(a) f_2 does not exist - only odd harmonics present with closed tube.

(b) $f_3 = 3 (343 \text{ m/s}) / [4 (0.80 \text{ m})]$; $f_3 = 322 \text{ Hz}$

83. $M = 1 / \sin\theta$; $2.0 = 1 / \sin\theta$; $\theta = 30°$

91. $\beta = 10 \log (I / I_o)$

$90\text{ dB} = 10 \log I / I_o$

$10^9 = I / (10^{-12}\text{ W/m}^2)$; $I = 10^{-3}\text{ W/m}^2$

$I = P / A$; $10^{-3}\text{ W/m}^2 = P / (1.5\text{ m}^2)$; $P = 1.5 \times 10^{-3}\text{ W}$

$P = E / t$; $1.5 \times 10^{-3}\text{ W} = E / (5.0\text{ s})$; $E = 7.5 \times 10^{-3}\text{ J}$

Sample Quiz

Use 340 m/s as the speed of sound .

Multiple Choice. Choose the correct answer.

__ 1. Thunder is heard 2.0 s after seeing a lighting bolt. Approximately how far was the bolt from the observer ?
A. 170 m B. 340 m C. 680 m D. 1360 m

__ 2. A pitch has a frequency of 4.0 MHz. What type of sound wave is it ?
A. audiable B. infrasonic C. ultrasonic

__ 3. The frequency of a note is 330 Hz. What is the frequency of the 2nd harmonic ?
A. 165 Hz B. 330 Hz C. 660 Hz D. 990 Hz

__ 4. Two tones have frequencies of 330 Hz and 332 Hz. What is the beat frequency?
A. 2 Hz B. 331 Hz C. 662 Hz D. 340 Hz

__ 5. A ship blows it fog horn and an echo is heard 4.0 s later by a person on the ship. How far is it from the ship to the point where the sound was reflected ?
A. 680 m B. 1020 m C. 1360 m D. 2720 m

__ 6. A person blows across the top of a bottle 20 cm tall which is open only on one end. What is the wavelengthof the fundamental ?
A. 5.0 cm B. 10 cm C. 40 cm D. 80 cm

__ 7. The intensity of a point source a distance d from the source is I. What is the intensity a distance d/2 from the source ?
A. I / 4 B. I / 2 C. 2 I D. 4 I

Problems

8. An ambulance driver is moving at 60 km/h and blows the siren which has a frequency 800 Hz. What is the frequency heard by
   (a) the ambulance driver ?
   (b) an observer as the ambulance appoaches ?
   (c) an observer in a car moving at 30 km/h in the same direction as the ambulance diver as the ambulance passes the observer ?

9. The sound level intensity for a jet is 110 dB. What would be the intensity for 3 similar jets ?

10. The frequency of the third harmonic of an closed organ pipe is 900 Hz. What is the length of the pipe ?

CHAPTER 15 Electric Charge, Force, and Energy

Chapter Objectives

Upon completion of the unit on electrostatics, the students should be able to :

1. compare and contrast gravitational and electrical forces.
2. state and apply the law of conservation of charge.
3. use the fundamental charge to determine the number of electric charges present.
4. distinguish between conductors and insulators.
5. explain the three ways to charge an object electrostatically.
6. state and apply Coulomb's law for electrostatic charges for forces in one and two dimensions.
7. find the magnitude and the direction of the electric field for a test charge, and for a point charge or combination of point charges.
8. describe the four conditions for electrostatic equlibrium.
9. find the electrostatic potential energy for two or more charges, and apply the law of conservation of energy to electrostatics.
10. state and apply electrostatic potential for test charges and for a point and group of point charges.
11. calculate the electric potential in a uniform electric field.
12. plot equipotential lines and lines of force for two charges.
13. recognize that electron volt (eV) is a unit for measuring energy, and state the definition for the eV.
14. work with capacitors in order to write expressions for and find the capacitance, charge, voltage, and energy for parallel plate capacitors.
15. explain and apply the concept of dielectrics.
16. find the capacitance, charge, and voltage for a group of capacitors connected in series, in parallel, and in a series-parallel combination of capacitors.
17. express a volt and a farad in base units.

Chapter Summary

***Electric charge** is a fundamental property. There are two types of charge, which are distinguished as being positive (+) or negative (-). A positive charge is arbitrarily associated with the proton, and a negative charge with the electron. The electronic charge is $e\pm = 1.6 \times 10^{-19}$ C, where the unit of charge is the coulomb (C).

*The directions of the electric forces on the charges of mutual interaction are given by the **law of charges**: like charges repel and unlike charges attract.

*Electrostatic charging can be done through charging by friction, charging by contact, and charging by induction.

*The magnitude of the electric force between two charges is given by **Coulomb's law** : $F = kq_1q_2/r^2$, where $k = 9.0 \times 10^9$ N-m^2/C^2. The **electric field** is the force per unit charge ($E = F/q_o$). A positive test charge is used to map an electric field and to give its direction.

*In electrostatic equilibrium :

(a) Any excess charge on an isolated conductor resides entirely on the surface of the conductor.
(b) The electric field is zero everywhere inside a charged conductor.
(c) The electric field at the surface of a charged conductor is perpendicular to the surface.
(d) Charge tends to accumulate at sharp points or locations of greatest curvature on asymmetric charged conductors.

*The mutual **electrostatic potential energy** of two charges is given by $U = kq_1q_2/r$. The signs (+ or -) of the charges mathematically determines whether the potential energy is positive or negative. The **electric potential** or **voltage** is the energy per unit charge, $V = U/q = kq/r$, and is given the unit of volt (V).

*The potential (voltage) difference is the work done in moving a test charge between two points, $V = V_B - V_A = -W_{AB} / qo$. In a uniform electric field, the potential difference in moving a charge through a straight-line distance d is $V = -Ed \cos \theta$, where θ is the angle between the electric field and the displacement. Equipotentials are lines in an electric field on which the potential is constant. These lines (surfaces in three dimensions) are at right angles to the electric fields lines, and no work is done in moving a charge along an equipotential.

*An **electron volt** (eV) is a unit of energy, and is defined as the energy acquired by an electron moving through a potential difference of one volt. (1 eV = 1.6×10^{-19} J)

***Capacitance** is the ratio of the charge and the voltage, $C = Q/V$, and has the unit of farad (F), The electric field between charges or charged conductors provided a method by which electrical energy can be stored. Parallel plates form a type of a capacitor. The energy stored in a capacitor is equal to $U = (1/2)CV^2$.

*An insulating material is called a dielectric, and in a capacitor, a dielectric increases the capacitance and the energy storage capacity of the capacitor. This capability varies with materials and is characterized by the **dielectric constant** K. With a dielectric constant between the plates of a capacitor, the capacitance increases, $C = KC_o$, where C_o is the capacitance without the dielectric (in a vacuum or in air).

As a result, the energy of the capacitor is $U = (1/2)CV^2 = (1/2)K(C_oV^2) = kU_o$.

*For capacitors in series, the charge on the capacitors is the same. For capacitors in parallel, the voltage across the capacitors is the same.

Important Terms and Relationships

15.1 Electric Charge

electric charge
law of charges
coulomb (C)
law of conservation of charge
quantization of electric charge : $q = ne$
$e = \pm 1.6 \times 10^{-19}$ C

15.2 Electrostatic Charging

conductors
insulators
semiconductors
electrostatic charging
charging by friction
charging by conctact
charging by induction
polarization

15.3 Electric Force

Coulomb's law $F = k\, q_1 q_2 / r^2$

15.4 Electric Field

electric field

electric field (point charge) $E = kq / r^2$

electric field (test charge) $E = F / q$

15.5 Electric Energy and Electric Potential

electrostatic (electric) potential energy : $U = kq_1q_2 / r$

electric potential (voltage , V)

electric potential (point charge) $V = kq / r$

potential difference : $\Delta V = V_B - V_A = -W_{AB} / q_o$

voltage in a uniform electric field : $V = -Ed \cos \theta$

volt (V)

equipotentials

electron volt (eV)

15.6 Capacitors and Dielectrics

capacitor

capacitance : $C = Q / C$

parallel plate capacitor : $C = \varepsilon_o A / d$

farad (F)

energy is a capacitor : $U = (1/2)CV^2 = (1/2)CV = Q^2 / 2C$

dielectric $V = V_o / K$; $C = KC_o$; $U = KU_o$

equivalent capacitance in series : $1/C_s = 1/C_1 + 1/C_2 + 1/C_3$

equivalent capcitance in parallel : $C_p = C_1 + C_2 + C_3 + ...$

Additional Solved Problems (The + charge is understood.)

15.1 Electric Charge

Example A piece of plastic has a charge of 4.8×10^{-17} C. How many more protons are there than electrons ?

given : $q = 4.8 \times 10^{-17}$ C

$q = ne$; $4.8 \times 10^{-17}\ C = n (1.6 \times 10^{-19}\ C)$

$n = 3.0 \times 10^2$ more protons than electrons

15.3 Electric Force

Example 1 A charge of 3.0 μC is placed at the 0 cm mark and a second charge of 12 μC is placed at the 100 cm mark on a meter stick. What is the magnitude of the force on the 12 μC charge ?

given : $q_1 = 3.0\ \mu C$; $q_2 = 12\ \mu C$; $r = 1.00$ m

$F = k q_1 q_2 / r^2$; $F = (9.0 \times 10^9\ N\text{-}m^2 / C^2)(3.0 \times 10^{-6}\ C)(12 \times 10^{-6}\ C) / (1.00\ m)^2$

$F = 0.32$ N

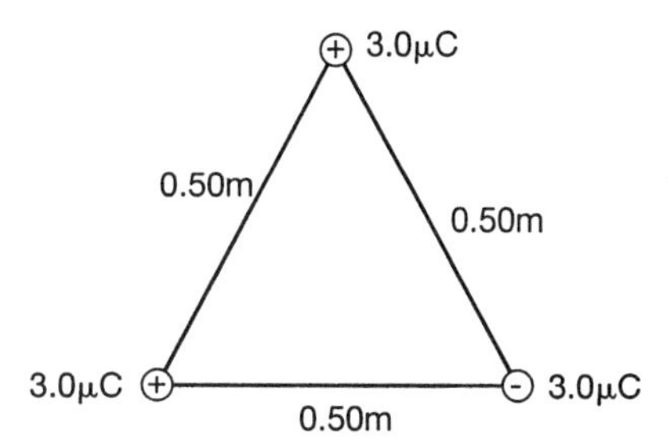

Fig. 15.1

Example 2 Three 3.0 μC charges are located at the vertices of an equliateral triangle as shown in Fig 15.1. What is the net force on the charge at the top of the triangle ? Each side of the triangle has a length of 0.50 m.

given : $q_1 = q_2 = -q_3 = 3.0\ \mu C$; $r = 0.50$ m

q_1 : top charge ; q_2 : charge on the left ; q_3 : charge on right

$F_{12} = F_{23} = F_{13} = k q_1 q_2 / r^2$;

$F = (9.0 \times 10^9\ N\text{-}m^2 / C^2)(3.0 \times 10^{-6}\ C)^2 / (0.50\ m)^2 = 0.32$ N

$\Sigma F_y = 0$ (by symmetry)

$\Sigma F_x = F_{12} (\cos 60°) + F_{13} (\cos 60°) = (0.32\ N)\cos 60° + (0.32\ N) \cos 60°$

$F_x = 0.32$ N in the +x direction

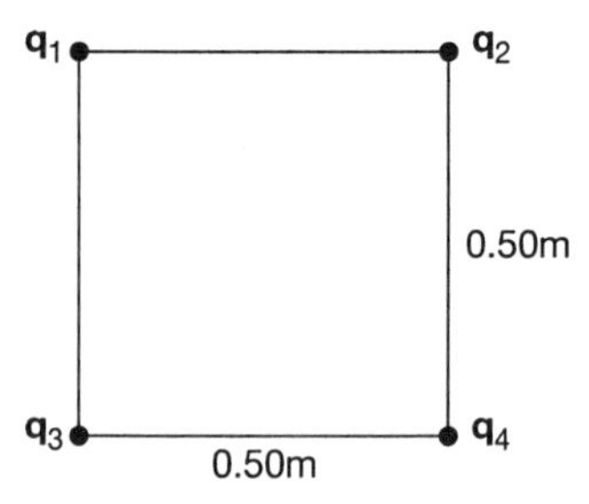

Fig. 15.2

Example 3 Four point 3.0 μC charges are located at the vertices of a square whose sides are 0.50 m (Fig. 15.2). Find the net force on the charge in the upper right corner.

Since adjacent charges and distances are the same as in the previous problem, the magnitude of the forces are the same. The force present between the two diagonally from each other are different since the distances are different.

$$r_{23} = [\,(0.50)^2 + (0.50)^2]^{1/2} = 0.71 \text{ m}$$

$$F_{23} = kq_2q_3 / r^2 = (9.0 \times 10^9 \text{ N-m}^2 / \text{C}^2)(3.0 \times 10^{-6} \text{ C})^2 / (0.71 \text{ m})^2$$

$$F_{23} = 0.16 \text{ N}$$

$$\Sigma F_x = F_{21} + F_{23} \cos 45° = 0.32 \text{ N} + (0.16 \text{ N})(\cos 45°) = 0.43 \text{ N}$$

$$\Sigma F_y = F_{24} + F_{23} \sin 45° = 0.32 \text{ N} + (0.16 \text{ N})(\sin 45°) = 0.43 \text{ N}$$

$F_2^2 = F_x^2 + F_y^2$; $F_2^2 = (0.43 \text{ N})^2 + (0.43 \text{ N})^2$; F = 0.61 N 45° above +x axis

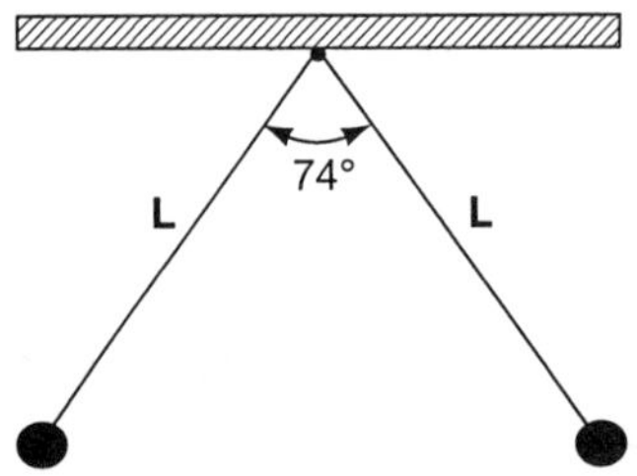

Fig. 15.3

Example 4 Two equally charged masses are suspended from light strings as shown in Fig. 15.3. The mass of each is 5.0 g and the angle between the two is 74°. The length of each string is 0.50 m.
A. What is the tension in the string ?
B. What is the electrostatic force present ?
C. What is the charge on each mass ?

given : $\theta = 74°$; $L = 0.50$ m ; $m = 5.0 \times 10^{-3}$ kg

the system is in static equilibrium or at rest :

A. $\Sigma F_x = 0 = T \sin 37° - F_e$ $\Sigma F_y = 0 = T \cos 37° - mg$

$0 = T\,(0.80) - (5.0 \times 10^{-3}\ \text{kg})(9.80\ \text{m/s}^2)$

$T = 0.61$ N

B. $0 = (0.61\ \text{N})(\sin 37°) - F_e$

$F_e = 0.38$ N

C. $F_e = k\,q_1 q_2 / r^2$ $r = 2\,(0.50\ \text{m})(\sin 37°) = 0.60$ m

$0.61\ \text{N} = (9.0 \times 10^9\ \text{N-m}^2 / \text{C}^2)(q^2) / (0.60\ \text{m})^2$; $q = 4.9 \times 10^{-6}$ C

15.4 Electric Field

Example 1 Find the electric field 0.50 m from a 5.0 μC point charge.

given : $q = 5.0$ μC ; $r = 0.50$ m

$E = kq / r^2$; $E = (9.0 \times 10^9\ \text{N-m}^2 / \text{C}^2)(5.0 \times 10^{-6}\ \text{C}) / (0.50\ \text{m})^2$

$E = 1.8 \times 10^5$ N/C away from the 5.0 μC charge

Example 2 A 5.0 μC point charge is placed at the 0 cm mark of a meterstick and a -4.0 μC charge is placed at the 50 cm mark.

A. What is the electric field at the 30 cm mark ?
B. At what point along the line connecting the two charges is the electric field zero ?

given : $q_1 = 5.0$ μC ; $q_2 = -4.0$ μC ; $x_1 = 0.30$ m ; $x_2 = 0.20$ m

A. $E = E_1 + E_2$ (add since vectors are in the same direction)

$E_1 = (9.0 \times 10^9\ \text{N-m}^2 / \text{C}^2)(5.0 \times 10^{-6}\ \text{C}) / (0.30\ \text{m})^2 = 5.0 \times 10^5$ N/C

$E_2 = (9.0 \times 10^9\ \text{N-m}^2 / \text{C}^2)(4.0 \times 10^{-6}\ \text{C}) / (0.20\ \text{m})^2 = 9.0 \times 10^5$ N/C

$E = 1.40 \times 10^6$ N/C toward the -4.0 μC charge.

B. $E_1 = E_2$

$k\, q_1 / (0.50\text{ m} + x)^2 = k\, q_2 / x^2$ (x as measured to the right of -4.0 μC charge)

$q_1^{1/2} / (0.50 + x) = q_2^{1/2} / x$; $5.0^{1/2} / (0.50 + x) = 4.0^{1/2} / x$

$(2.2) / (0.50 + x) = (2.0) / x$; $2.2\, x = 1.0 + 2.0\, x$; $0.20\, x = 1.0$; $x = 5.0$ m from the 4.0 μC charge or 5.50 m from the 0 cm mark of the meterstick.

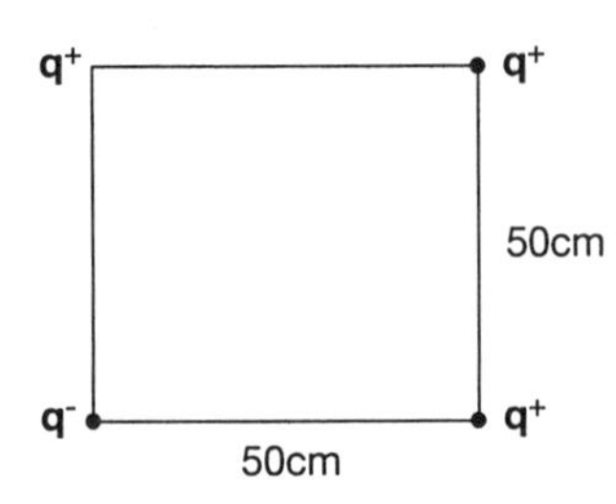

Fig. 15.4

Example 3 Find the electric field at the center of the square shown in Fig 15.4. Each charge has a magnitude of 3.0 μC and the measure of each side is 50 cm.

given : $q = 3.0$ μC ; $r = 0.50$ m

To make the problem easier, rotate the x - y coordinate plane 45° in the clockwise direction. To find r take half the diagonal : 0.35 m

$\Sigma E_x = E_1 - E_4 = 0$ (since $E_1 = E_4$)

$\Sigma E_y = E_2 + E_3 = 2\,(9.0 \times 10^9\text{ N-m}^2 / C^2)\,(3.0 \times 10^{-6}) / (0.35\text{ m})^2 = 4.4 \times 10^5$ N/C toward the negative charge.

Example 4 A -4.0 μC experiences an electric field of 5.0×10^5 N/C in the +x direction. What is the magnitude and the direction of the electrical force acting on the mass ?

given : $q = -4.0$ μC ; $E = 5.0 \times 10^5$ N/C in +x

$E = F / q$; $5.0 \times 10^5\text{ N/C} = F / (-4.0 \times 10^{-6}\text{ C})$; $F = 2.0$ N

since the charge is negative, the force and the field are in opposite directions of each other : 2.0 N in the - x direction.

Example 5 A proton initially at rest is accelerated in an electric field of 1.0×10^3 N/C for 1.0 μs.
A. What is the acceleration of the proton ?
B. What is the speed of the proton after period of acceleration ?
C. What distance does the proton travel during this period of time ?

given : $q = 1.6 \times 10^{-19}$ C ; $m = 1.67 \times 10^{-27}$ kg ; $E = 1.0 \times 10^3$ N/C
$t = 1.0 \times 10^{-6}$ s ; $v_o = 0$

A. $E = F / q$; 1.0×10^3 N/C $= F / (1.6 \times 10^{-19}$ C) ; $F = 1.6 \times 10^{-16}$ N
$F = ma$; 1.6×10^{-16} N $= (1.67 \times 10^{-27}$ kg)(a) ; $a = 9.6 \times 10^{10}$ m/s^2

B. $v = v_o + at$; $v = 0 + (9.6 \times 10^{10}$ m/s^2)(1.0×10^{-6} s) $= 9.6 \times 10^4$ m/s

C. $x = v_o t + (1/2)at^2$; $x = 0 + (1/2)(9.6 \times 10^{10}$ m/s^2)(1.0×10^{-6} s)^2
$x = 4.8 \times 10^{-2}$ m

15.5 Electric Energy and Electric Potential

Example 1 A proton, initially at rest, is subjected to an electric potential of 500 V.
A. What is the kinetic energy of the proton ?
B. What is the speed of the proton ?

given : $m = 1.67 \times 10^{-27}$ kg ; $q = 1.6 \times 10^{-19}$ C ; $V = 500$ V

A. $V = K / q$; 500 V $= K / (1.6 \times 10^{-19}$ C) ; $K = 8.0 \times 10^{-17}$ J

B. $K = (1/2)mv^2$; 8.0×10^{-17} J $= (1/2)(1.67 \times 10^{-27}$ kg) v^2
$v = 3.1 \times 10^5$ m/s

Example 2 An electric field of 1.0×10^3 N/C exists between two parallel plates which are separated by a distance of 10 cm. What is the potential difference between the two plates ?

given : $E = 1.0 \times 10^3$ N/C ; $d = 0.10$ m
$V = Ed$
$V = (1.0 \times 10^3$ N/C)(0.10 m) $= 1.0 \times 10^2$ V

Example 3 A 5.0 μC point charge is located at the 0 cm mark and a second charge -10.0 μC is located at x = 60 cm.
A. At what point is the electric potential zero ?
B. What is the potential energy of the system ?

given : $q_1 = 5.0\ \mu C$; $q_2 = -10\ \mu C$; $x_1 = 0$; $x_2 = 60$ cm

A. $0 = V_1 + V_2$

$0 = kq_1 / r_1 + kq_2 / r_2$

$-kq_2 / r_2 = kq_1 / r_1$

$- k (-10\ \mu C) / (60\ cm - x) = k (5.0\ \mu C) / x$

$60 - x = 2x$

$60 = 3x$

$x = 20$ cm

B. $U = kq_1q_2 / r$

$U = (9.0 \times 10^9\ N\text{-}m^2 / C^2)(5.0 \times 10^{-6}\ C)(-10 \times 10^{-6}\ C) / 0.60\ m = -0.75\ J$

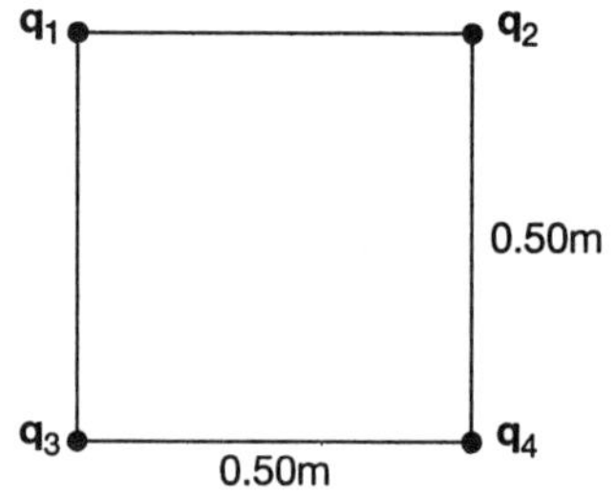

Fig. 15.4

Example 4 Find the electric potential at the center of the square shown in Fig. 15.4. Each charge has a magnitude of 3.0 μC and each side has a length of 50 cm. (q_1 q_2, and q_4 are positive and q_3 is negative)

given : $q_1 = q_2 = -q_3 = q_4 = 3.0\ \mu C$; $r = 0.50$ m

$V = V_1 + V_2 + V_3 + V_4$

$V_1 = kq_1 / r = (9.0 \times 10^9\ N\text{-}m^2 / C^2)(3.0 \times 10^{-6}\ C) / (0.35\ m) = 7.7 \times 10^4\ V$

$V = V_1 + V_2 = 2\ (7.7 \times 10^4\ V) = 1.5 \times 10^5\ V$

Example 5 Two 3.0-g point charges have equal and opposite charges of 3.0 mC. Initially the charges are separated by a distance of 2.0 m. What is the the speed of each charge after each has moved 20 cm ?

given : $m_1 = m_2 = 3.0 \times 10^{-3}$ kg ; $q_1 = -q_2 = 3.0 \times 10^{-3}$ C

$r_1 = 2.0$ m ; $r_2 = 1.6$ m

use the law of conservation of energy :

$U_o = U + K$

$U_o = kq_1q_2/r_1 = (9.0 \times 10^9 \text{ N-m}^2/\text{C}^2)(3.0 \times 10^{-3}\text{ C})(-3.0 \times 10^{-3}\text{ C}) / (2.0\text{ m})$

$U_o = -4.1 \times 10^4$ J

$U = (9.0 \times 10^9 \text{ N-m}^2/\text{C}^2)(3.0 \times 10^{-3}\text{ C})(-3.0 \times 10^{-3}\text{ C}) / (1.6\text{ m})$

$U = -5.1 \times 10^4$ J

$(-4.1 \times 10^4\text{ J}) = (-5.1 \times 10^4\text{ J}) + 2K$

$K = 5.0 \times 10^3$ J

$K = (1/2)\,mv^2$; $5.0 \times 10^3 = (1/2)(3.0 \times 10^{-6}\text{ kg})(v^2)$

$v = 5.8 \times 10^4$ m/s

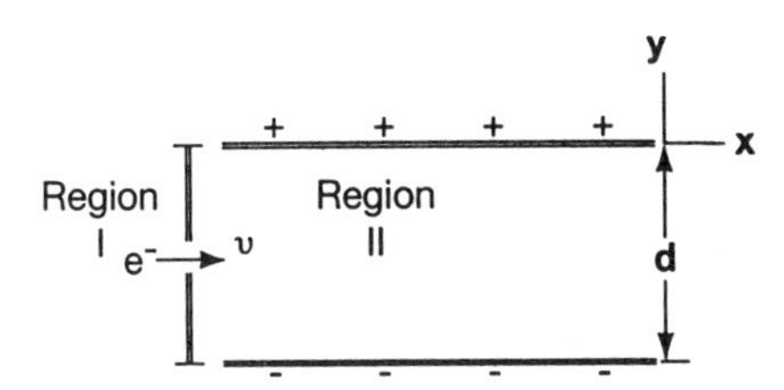

Example 6 An electon is accelerated by a gun in region I (to the left) and aquires a velocity **v** when it leaves region I. It then passes through region II (left) which has an electric field **E** directed in the -y direction. Let **y** be the distance between the plates in region II. In terms of given, information, m (mass of electron), e (charge of an electron), d (the horizontal distance traveled), find expression for the following.

A. The electric potential in region I.
B. The electric poetntial in region II.
C. The time the mass is in region II.
D. The vertical distance the mass travels in region II.
E. Sketch on the diagram the path traveled by the charge in region II.

A. $K = (1/2)mv^2$; $V = K / q$; $V = (1/2)\ mv^2 / e$; $V = mv^2 / 2e$
B. $V = Ed$; $V = Ed$
C. $x = v_x t$; $t = x / v_x$; $t = x / v_x$
D. $y = v_{oy}t - (1/2)at^2$; $Eq = ma$; $a = Eq / m$
$y = (1/2)(Eq/m)\ t^2$; $y = Eqt^2 / 2m$
E. The mass would travel in a parabolic path which is concave up.

15.6 Capacitors and Dielectrics

Example 1 A parallel plate capacitor consists of two plates separated by air. The plates have an area of $1.5 \times 10^{-4}\ m^2$ and are separated by a distance of 1.0 mm. The capacitor is then connected to a 6.0 V battery.
A. What is the capacitance ?
B. What is the charge on the plates ?
C. How much energy is stored in the electric field ?
D. What is the electric field between the plates ?

given : $\varepsilon_o = 8.85 \times 10^{-12}$ F/m ; $A = 1.5 \times 10^{-4}\ m^2$; $d = 1.0 \times 10^{-3}$ m ; $V = 6.0$ V

A. $C = \varepsilon_o A / d$; $C = (8.85 \times 10^{-12}\ F/m)(1.5 \times 10^{-4}\ m^2) / (1.0 \times 10^{-3}\ m)$
$C = 1.3 \times 10^{-12}$ F
B. $C = Q / V$; $1.3 \times 10^{-12}\ F = Q / (6.0\ v)$; $Q = 7.8 \times 10^{-12}$ C
C. $U = (1/2)CV^2$; $U = (1/2)(1.3 \times 10^{-12}\ F)(6.0\ V)^2 = 2.3 \times 10^{-11}$ J
D. $V = Ed$; $6.0\ V = E\ (1.0 \times 10^{-3}\ m)$; $E = 6.0 \times 10^3$ V

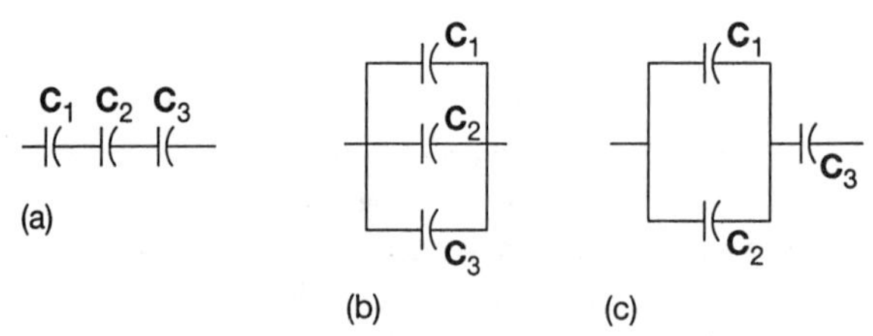

Fig. 15.6

Example 2 Find the equivalent capacitance in each diagram shown in Fig. 15.6 with $C_1 = 6.0\ \mu F$, $C_2 = 12\ \mu F$, $C_3 = 4.0\ \mu F$.

(a) $1/C = 1/C_1 + 1/C_2 + 1/C_3$
$1/C = 1/6\ \mu F + 1/12\ \mu F + 1/4.0\ \mu F$
$C = 2.0\ \mu F$

(b) $C = C_1 + C_2 + C_3$
$C = 6.0\ \mu F + 12\ \mu F + 4.0\ \mu F = 22\ \mu F$

(c) $C_p = C_1 + C_2$; $C_p = 6.0\ \mu F + 12\ \mu F = 18\ \mu F$
$1/C_t = 1/C_p + 1/C_3$
$1/C_t = 1/18\ \mu F + 1/4.0\ \mu F$
$C_t = 3.3\ \mu F$

Solutions to paired problems and additional solved problems

1. each proton and electron has a charge of $\pm 1.6 \times 10^{-19}$ C ;
two electrons (-3.2×10^{-19} C) + one proton (1.6×10^{-19}C) = -1.6×10^{-19} C

6. (a) by the law of conservation of charge, if the rod has a charge of $+8.0 \times 10^{-10}$ C, then the silk has a charge of -8.0×10^{-19} C .
(b) $q = ne$
$8.0 \times 10^{-10} = n\,(1.6 \times 10^{-19})$; $n = 5.0 \times 10^{9}$ electrons

11. After the electroscope is grounded, a positive charged rod could be brought close to it, causing electrons to move from the ground to the electroscope. To prove it was negatively charged, bring a charged rod (either positive or negative) could be brought close to the bulb. A negatively charged rod would cause the leaves to separate further, and a positively charged rod would cause them to come closer together.

15. $F = kq_1q_2 / r^2$;
(a) if the distance were halved the demoninator would change by $(1/2)^2$ or 1/4. Since the denominator decreases, the force increases. $F = (4)\ F_o$
(b) if the distance is tripled, the force decreases by a factor of (1/9).

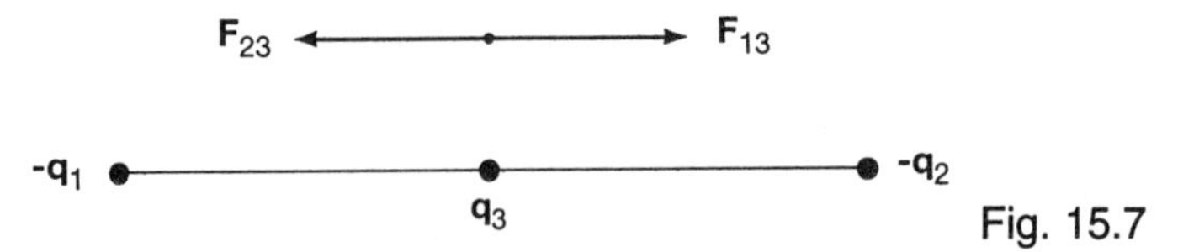

Fig. 15.7

20. (a) As shown in Fig. 15.7, the charge at the center would experience two forces. Since the forces are equal and opposite, the net force is zero.
(b) the only difference is that the forces are in opposite directions of each other.

24. (a) $F = kq_1q_2 / r^2$;
$F = (9.0 \times 10^9 \text{ N-m}^2 / C^2)(-1.6 \times 10^{-19} \text{ C})(1.6 \times 10^{-19} \text{ C}) / (5.3 \times 10^{-11} \text{ m})^2$
$F = 8.2 \times 10^{-8}$ N (attractive force)
(b) $F = mv^2 / r$; $8.2 \times 10^{-8} \text{ N} = (9.1 \times 10^{-31} \text{ kg})\, v^2 / (5.3 \times 10^{-11} \text{ m})$
$v = 2.2 \times 10^6$ m/s

30. (d) a volt is a J/C ; a J = N-m ; N-m / C-m = N / C

38. $E = kQ / r^2$;
$E_1 = (9.0 \times 10^9 \text{ N-m}^2 / C^2)(4.0 \times 10^{-6} \text{ C}) / (0.10 \text{ m})^2 = 3.6 \times 10^6$ N/C (- x)
$E_2 = (9.0 \times 10^9 \text{ N-m}^2 / C^2)(5.0 \times 10^{-6} \text{ C}) / (0.10 \text{ m})^2 = 4.5 \times 10^6$ N/C (+x)
$E_{net} = E_2 - E_1 = (4.5 \times 10^6 \text{ N/C}) - (3.6 \times 10^6 \text{ N/C})$
$E_{net} = 9.0 \times 10^5$ N/C toward -5.0 μC charge

44. rotate the x and y axis so the coordinate axis is along the diagonal of the square.
$E_1 = kq_1 / r^2$; $E_1 = (9.0 \times 10^9 \text{ N-m}^2 / C^2)(10 \times 10^{-6} \text{ C}) / (0.71 \text{ m})^2$
$E_1 = 1.8 \times 10^7$ N/C
$E_3 = (9.0 \times 10^9 \text{ N-m}^2 / C^2)(10 \times 10^{-6} \text{ C}) / [(0.71 \text{ m})(1.4)]^2 = 9.0 \times 10^6$ N/C
$E_x = 1.8 \times 10^7 \text{ N/C} + (0.9 \times 10^7 \text{ N/C}) = 2.7 \times 10^7$ N/C
$E_y = 1.8 \times 10^7 \text{ N/C} + (0.9 \times 10^7 \text{ N/C}) = 2.7 \times 10^7$ N/C
$E^2 = E_x^2 + E_y^2$; $E = 3.8 \times 10^7$ N/C in the +y direction

49. (c) the energy per unit charge

54. (a) $U_o = kq_1q_2 / r$;

$U_o = (9.0 \times 10^9 \text{ N-m}^2 / C^2)(-4.0 \times 10^{-6} \text{ C})(6.0 \times 10^{-6} \text{ C}) / (0.40 \text{ m}) = 0.54 \text{ J}$

$U = (9.0 \times 10^9 \text{ N-m}^2 / C^2)(-4.0 \times 10^{-6} \text{ C})(-6.0 \times 10^{-6} \text{ C}) / (0.40 \text{ m}) = 0.24 \text{ J}$

$\Delta U = U - U_o = -0.30 \text{ J}$

(b) no ; conservative system

57. $U = U_{12} + U_{23} + U_{13}$ (scalars) ; $U = kq_1q_2 / r$

$U_{12} = (9.0 \times 10^9 \text{ N-m}^2 / C^2)(4.0 \times 10^{-6} \text{ C})(4.0 \times 10^{-6} \text{ C}) / (0.20 \text{ m}) = 0.72 \text{ J}$

$U_{23} = (9.0 \times 10^9 \text{ N-m}^2 / C^2)(4.0 \times 10^{-6} \text{ C})(-4.0 \times 10^{-6} \text{ C}) / (0.20 \text{ m}) = -0.72 \text{ J}$

$U_{13} = (9.0 \times 10^9 \text{ N-m}^2 / C^2)(4.0 \times 10^{-6})(-4.0 \times 10^{-6}) / (0.20 \text{ m}) = -0.72 \text{ J}$

$U = (0.72 \text{ J}) + (-0.72 \text{ J}) + (-0.72 \text{ J}) = -0.72 \text{ J}$

63. $U_1 + K_1 = U_2 + K_2$

$(1/2)(1.67 \times 10^{-27} \text{ kg})(0.50 \text{ m/s})^2 + (9.0 \times 10^9 \text{ N-m}^2/C^2)(1.6 \times 10^{-19} \text{ C})^2/(1.0 \text{ m})$

$= U_2 + 0$; $U_2 = 4.4 \times 10^{-28} \text{ J}$

$U_2 = k\, q_1q_2 / r_2$; $4.4 \times 10^{-28} \text{ J} = (9.0 \times 10^9 \text{ N-m}^2 / C^2)(1.6 \times 10^{-19})^2 / r_2$

$r_2 = 0.52 \text{ m}$

68. (a) $V = Ed$; $V = (2.0 \times 10 \text{ N/C})(0.050 \text{ m}) = 3.0 \times 10^4 \text{ V}$

$\Delta V = -W / q$; $3.0 \times 10^4 \text{ V} = -W / (1.6 \times 10^{-19} \text{C})$

against field : -4.8×10^{-15} J

with the field : 4.8×10^{-15} J

(b) $q = e$; against : -3.0×10^4 eV or -30 keV

with the field : 3.0×10^4 eV or -30 keV

(c) zero , since $W = Fd \cos \theta$; $\cos 90 = 0$, therefore work equals zero

75. The charged rubber rod repels the polar water molecules. A nonpolar liquid stream would not be deflected. On a humid day, electrostatic charging is difficult because there is a thin film of water on the objects and the charge is conducted away, preventing a buildup.

80. (a) $C = \varepsilon_o A / d$; $C = (8.85 \times 10^{-12}\ F/m)(0.20\ m^2) / (5.0 \times 10^{-3}\ m) = 3.5 \times 10^{-10}\ F$

$C = Q / V$; $3.5 \times 10^{-10}\ F = Q / (12\ V)$; $Q = 4.2 \times 10^{-9}\ C$

(b) $U = (1/2)CV^2$; $U = (1/2)(3.5 \times 10^{-10}\ F)(12\ V)^2 = 2.5 \times 10^{-8}\ J$

85. (a) in series : $(1/C_s) = (1/C_1) + (1/C_2)$

$(1/C_s) = (1/0.60\ \mu F) + (1/0.80\ \mu F)$; $C_s = 0.34\ \mu F$

(b) in parallel : $C_p = C_1 + C_2$

$C_p = (0.60\ \mu F) + (0.80\ \mu F) = 1.40\ \mu F$

89. $C_{parallel} = C_1 + C_2$

$(1 / C_t) = (1 / C_3) + (1 / C_{eq})$

$(1 / (0.75\ \mu F) = (1 / 1.0\ \mu F) + 1 / C_{eq}$; $C_{parallel} = 3.0\ \mu F$

$C_{parallel} = C_1 + C_2$; $3.0\ \mu F = 1.0\ \mu F + C_2$; $C_2 = 2.0\ \mu F$

95. when the charges are located at an infinite distance from each other, the potential energy is zero.

now find the new potential energy : $U = kq_1q_2 / r$

$U = (9.0 \times 10^9\ N\text{-}m^2 / C^2)(+2.5 \times 10^{-6}\ C)(4.5 \times 10^{-6}\ C) / (0.50\ m)$

$U = 0.20\ J$

$W = 0.20\ J$

Chapter 15

Sample Quiz

Multiple Choice. Choose the correct answer.

__ 1. Two point charges have a force F between them when they are a distance d apart. The distance now is decreased to d / 2. What is the new force ?
A. F /4 B. F / 2 C. (2) F D. 4 F

__ 2. A 0.20 C charge is moved from a potential of 10 V to a potential of 30 V. What is the work done ?
A. 6.0 J B. 4.0 V C. 0.033 J D. 0.010 J

__ 3. A parallel plate capacitor has a capacitance C. The area of the plates is doubled and the distance between the plates is doubled. What is the new capacitance ?
A. C / 4 B. C / 2 C. C D. 4C

__ 4. A 2.0 μF capacitor is charged to an electric potential of 12 V. What is the charge stored in the capacitor ?
A. 6.0 μC B. 24 μC C. 0.17 μC D. 48 μC

__ 5. What is the energy stored in the electric field question 4 ?
A. 72 μJ B. 144 μJ C. 288 μJ D. 576 μJ

__ 6. Two parallel plates have an electric potential of 50 V and are separated by a distance of 10 cm. What is the electric field between the plates ?
A. 500 N/C B. 5 N/C C. 5000 N/C D. 50 N/C

__ 7. Which of the following is not a vector ?
A. electric force B. acceleration C. electric potential D. electric field

Problems

8. Three point charges are located at the following positions
$q_1 = 2.0\ \mu C$ at $x = 1.0$ m
$q_2 = 3.0\ \mu C$ at $x = 0$ m
$q_3 = -4.0\ \mu C$ at $x = -1.0$ m
What is the force on the 3.0 μC charge ?

9. Two 3.0-mg, 3.0-μC point charges are located on the ends of a meterstick.
Find the following for center of the meterstick
A. the electric field.
B. the electric potential

10. A parallel plate capacitor has plates with area of 0.050 m^2 and separated with air by a distance of 2.0 mm. The capacitor is charged by a 12 V battery.
A. What is the capacitance ?
B. What is the charge stored on the plates ?
C. What is the energy stored in the electric field ?

CHAPTER 16 Electric Current and Resistance

Chapter Objectives

Upon completion of the unit on electric current and resistance, students should be able to :

1. describe the parts of a battery.
2. describe the concept of electromotive force.
3. explain electric current and state the direction of electron flow and the direction of conventional current, and relate the flow of electric current to the rate of flow of charge.
4. write an ampere and an ohm in basic units.
5. describe drift velocity.
6. state and apply Ohm's law, and recognize when Ohm's law is not obeyed.
7. describe how the resistance depends upon area, temperature, length, and resistivity.
8. write an expression, and apply the expression, for the resistance of a conductor in terms of the resistivity, length, and area.
9. write an expression and apply the expression for the change in resistivity and the resistance of a conductor as they vary with respect to the temperature.
10. relate conductivity to resistance.
11. relate the current, the resistance, and the voltage for a circuit to the electric power.
12. calculate the cost of electrical energy.

Chapter Summary

* A battery is a device that converts chemical energy into electrical energy through the chemical action of the metal electrodes in an electrolyte. The potential difference across the terminals of a battery when not connected to an external circuit is called the **electromotive force** (emf). The operating voltage of a battery is less than the emf because of internal resistance.

***Direct current (dc)** flows only in one direction in a circuit, i.e., from the negative (-) terminal to the positive (+) terminal for electron flow in a battery circuit.

* **Electric current** (I) in a conductor is defined as the net amount of charge passing through a cross-sectional area of the conductor per unit time , $I = q/t$. The unit of current is coulomb/s, which is given the name of **ampere** (A) are "amp."

*A sustained electric current requires a voltage source and a complete circuit. Circuit analysis is usually done in terms of the conventional current, which is in the direction positive charge would flow, or opposite to the electron flow. The net electron flow, in a conductor is characterized by the average drift velocity.

***Ohm's law, $V = IR$,** applies to many but not all materials and provides a definition for resistance. For an ohmic material, R is constant (at a given temperature).

*The major factors affecting the resistance of a conductor of uniform cross-section are:
(1) the type of material
(2) its length
(3) its cross-sectional area
(4) its temperature

*The resistive property of a particular material is characterized by its **resistivity**, where for a conductor of uniform cross-section, $\rho = RA/L$. The resistivity is somewhat temperature dependent, and over a small range of temperature change (ΔT), the resistivity changes according to $\rho = \rho_o (1 + \alpha\Delta T)$, where ρ_o is some reference resistivity at T_o and α is the **temperature coefficient of resistivity**. The reciprocal of the resistivity is called the conductivity, $\sigma = 1/\rho$.

*For a conductor of uniform cross-section, we may write the variation of resistance with temperature as $R = Ro (1 + \alpha\Delta T)$. Most materials have positive α's and their resistance increases with increasing temperature.

*Electric power is given by $P = IV = V^2/R = I^2R$. The energy dissipated in a current-carrying conductor is sometimes referred to a **joule heat** or I^2R loss.

Important Terms and Relationships

16.1 Batteries and Direct Current

electric current
battery
cathode
anode
electromotive force (emf)
terminal voltage

16.2 Current and Drift Velocity

conventional current
current : $I = q / t$
ampere (A)
drift velocity

16.3 Ohm's Law and Resistance

resistance
Ohm's law $V = IR$
ohm (Ω)
resistivity $\rho = RA / L$
conductivity $\sigma = 1/\rho$
temperature coefficient of resistivity : $\rho = \rho_o = (1 + \alpha \, \Delta T)$
temperature-dependence on resistance : $R = R_o \;\alpha \, \Delta T$
superconductivity

16.4 Electric Power

electric power $P = IV$; $P = I^2R$; $P = V^2 / R$
joule heat $P = I^2R$

<u>Additional Solved Problems</u>

16.1 Batteries and Direct Current

Example Three 6.0 V batteries are connected together.
A. What is the total voltage if the cells are connected in parallel ?
B. What is the total voltage if the cells are connected in series ?

A. when cells are connected in parallel their voltage is constant :
$V_t = V_1 = V_2 = V_3 = 6.0 \text{ V}$

B. when cells are connected in series, their voltage is equal to the sum :
$V_t = V_1 + V_2 + V_3 = 6.0 \text{ V} + 6.0 \text{ V} + 6.0 \text{ V} = 18.0 \text{ V}$

16.2 Current and Drift Velocity

Example 1 A charge of 500 C flows through a wire in 2.0 minutes. What is the current ?

$q = It$; $q = 500$ C ; $t = 2.0$ min $= 120$ s
500 C $= I$ (120 s) ; $I = 4.2$ A

Example 2 A current of 3.5 A flows through a wire for 5.0 minutes.
A. How much current flows in the wire during this time ?
B. How many electrons flow during this time ?

A. $q = It$; $t = 5.0$ min $= 300$ s
$q = (3.5 \text{ A})(300 \text{ s}) = 1.1 \times 10^{3}$ C

B. $(1.1 \times 10^{3} \text{ C}) = n\,(1.6 \times 10^{-19} \text{ C})$; $n = 6.9 \times 10^{21}$ electrons

16.3 Ohm's Law and Resistance

Example 1 A hair dryer is connected to a 120 V source with a current 10 A. What is the resistance of the dryer ?

given : $V = 120$ V ; $I = 10$ A
$V = IR$
$120 \text{ V} = (10 \text{ A}) R$; $R = 12\ \Omega$

Example 2 A copper wire has a length of 10.0 m and a cross-sectional radius of 1.0 mm. What is the resistance of the wire ?

given : $\rho = 1.70 \times 10^{-8}\ \Omega\text{-m}$; $L = 10.0$ m ; $r = 1.0 \times 10^{-3}$ m
$R = \rho L / A$; $R = (1.70 \times 10^{-8}\ \Omega\text{-m})(10.0 \text{ m}) / [\pi\,(1.0 \times 10^{-3} \text{ m})^2]$
$R = 5.4 \times 10^{-2}\ \Omega$

Example 3 An 80.0-m length aluminum wire has a resistance of 1.0 Ω. What is the diameter of the wire ?

given : $\rho = 2.82 \times 10^{-8}$ Ω-m ; L = 80.0 m ; R = 1.0 Ω

$R = \rho L / A$; $1.0\ \Omega = (2.82 \times 10^{-8}\ \Omega\text{-m})(80.0\ \text{m}) / A$

$A = 2.3 \times 10^{-6}\ \text{m}^2$; $A = \pi d^2 / 4$; $2.3 \times 10^{-6}\ \text{m}^2 = \pi d^2 / 4$

$d = 1.7 \times 10^{-3}$ m

Example 4 A carbon resistor has a resistance of 20 Ω at a temperature of 20°C. What is the resistance if heats up to a temperature 150°C ?

given : $R_o = 20\ \Omega$; $\Delta T = 130$ C° ; $\alpha = -0.005\ \text{C}^{-1}$

$R = R_o (1 + \alpha \Delta T) = (20\ \Omega) [1 + (-0.005\ \text{C}^{-1})(130\ \text{C}°) = 7.0\ \Omega$

16.4 Electric Power

Example 1 A 100 W light bulb is connected to 120 V outlet for 2.0 h.

A. What is the current in the circuit ?
B. What is the resistance of the bulb ?
C. What is the amount of electrical energy produced in this time ?
D. How much will it cost for operating the bulb if electricity costs $0.12 per kWh ?

given : P = 100 W ; V = 120 V ; t = 2.0 h ; cost $0.12 1/ kW-h

A. $P = IV$
$100\ \text{W} = i\ (120\ \text{V})$; $I = 0.83$ A

B. $V = IR$
$120\ \text{V} = (0.83\ \text{A})\ R$; $R = 145\ \Omega$

C. $P = E / t$; $100\ \text{W} = E / (7.2 \times 10^3\ \text{s})$; $E = 7.2 \times 10^5$ J

D. (0.100 kW)(2.0 h) $0.12 / kWh = 2.4¢

Example 2 An large electric heater is rated at 50 kW at 220 V.

A. What is the electric current ?
B. What is the electric resistance ?
C. What is the amount of electrical energy produced in 2.0 h ?
D. How much does it cost for operating the heater for this time if electricity costs $0.10 per kW-h ?

given : $P = 50 \times 10^3$ W ; V = 220 V ; t = 20 h ; cost $0.10 1 / kWh

A. $P = IV$

$(50 \times 10^3 \text{ W}) = I\,(220 \text{ V})$; $I = 2.3 \times 10^2$ A

B. $V = IR$

$220 \text{ V} = (2.3 \times 10^2 \text{ A})\,R$; $R = 0.96\ \Omega$

C. $P = E / t$

$50 \times 10^3 \text{ W} = E / (7200 \text{ s})$; $E = 3.6 \times 10^8$ J

D. (50 kW)(2.0 h) ($0.10) = $10

Example 3 Two kilograms of water are heated from room temperature to 80°C in 10 minutes by electric immersion heater. The applied voltage is 24 V.

A. What is the electric resistance ?
B. What is the current in the heater ?

given : m = 2.0 kg ; c = 4190 J / kg-C° ; ΔT = 60 C° ; t = 600 s ; V = 24 V

A. first find the heat : $Q = mc\Delta T$;

$Q = (2.0 \text{ kg})(4190 \text{ J/kg-C°})(60 \text{ C°}) = 5.0 \times 10^5$ J

now find the power

$P = E / t$; $P = 5.0 \times 10^5 \text{ J} / 600 \text{ s} = 8.3 \times 10^2$ W

now find the resistance :

$P = V^2 / R$

$8.3 \times 10^2 \text{ W} = (24 \text{ V})^2 / R$; $R = 0.69\ \Omega$

B. $V = IR$

$24 \text{ V} = I\ (0.69\ \Omega)$; $I = 35$ A

Solutions to paired problems and other selected problems

4. The terminal voltage is the difference between the emf and the product of the current and the internal resistance. The terminal voltage is then the voltage across the external circuit. For an open circuit, there is no current flowing in the circuit and the emf is the terminal voltage.

10. $q = It$; $1.8\ C = (3.0 \times 10^{-3}\ A)\ t$; $t = 6.0 \times 10^{2}\ s = 10\ min$

16. (a) ΔQ = (number of charges)(charge per particle) = $(nAx)(q)$
(b) $x = V_d \Delta t$ and $\Delta Q = (nAv_d \Delta t)\ q$; $I = \Delta Q / \Delta t$
$I = nqV_dA$

20. The slope of a V versus current graph gives the resistance of resistor. Since the slopes are different, the resistances are different.

27. two batteries in series : $V_s = V_1 + V_2 = (9.0\ V) + (9.0\ V) = 18.0\ V$
$V = IR$; $18.0\ V = I\ (100\ \Omega)$; $I = 0.18\ A$

30. $R_s / R_L = L_s / L_L = (0.50\ m) / (2.0\ m) = 0.25$

34. $\Delta R = \alpha R_o \Delta T$; $\Delta R = (6.80 \times 10^{-3}\ C^{\circ -1})(2.50 \times 10^{-2}\ \Omega)(27\ C^{\circ})$; $\Delta R = 4.6\ m\Omega$

43. $V = I_o R_o$; $6.0 = (0.50)\ R_o$; $R_o = 12\ \Omega$
$R = R_o\ (1 + \alpha \Delta T)$; $R = 12\ \Omega\ [\ 1 + (-0.07\ C^{\circ -1})(5.0\ C^{\circ})$; $R = 7.8\ \Omega$
$6.0\ V = I\ (7.8\ \Omega)$; $I = 0.77\ A$

48. $P = V^2 / R$; if the voltage is doubled, the power is quadrupled.

58. (1.500 kW)(10 min)(1 h / 60 min)(30)(12¢) = 90¢

65. $P = V^2 / R$; $150\ W = (120\ V)^2 / R$; $R = 96\ \Omega$

69. first find the power : $P = V^2 / R$; $P = (120\ V)^2 / (9.0\ \Omega) = 1.6 \times 10^3\ W$
next find the heat needed to change the water to its boiling point :
$Q = mc\Delta T = (0.25\ kg)(4190\ J/kg\text{-}C°)(80\ C°) = 8.4 \times 10^4\ J$
$P = E / t$; $1.6 \times 10^3\ W = (8.4 \times 10^4\ J) / t$; $t = 53\ s$

Sample Quiz

Multiple Choice. Choose the correct aswer.

__ 1. A resistor has a current of 2.0 A and a voltage of 50 V. What is the resistance ?
A. 0.040 Ω B. 25 Ω C. 100 Ω D. 400 Ω

__ 2. Which of the following is the best conductor of electricity ?
A. aluminum B. copper C. steel D. iron

__ 3. Which of the following combinations of the length and cross-sectional area for a wire has the least resistance ?
A. L and A B. 2L and (1/2) A C. (1/2)L and 2 A D. L and (1/2)A

__ 4. During a power demand the voltage output is reduced by 10%. By what percent is the power of a resistor affected ?
A. 3.2% less B. 10% less C. 19% less D. 81% less

__ 5. How much does is cost to operate a 100 W light bulb (rated at 120 V) for a week if electricity costs $0.10 per kWh ?
A. $1680 B. $16.80 C. $1.68 D. $0.17

__ 6. The length of a wire is doubled and the radius is doubled. By what factor does the resistance change ?
A. 4 time larger B. twice as large C. unchanged D. half as large

__ 7. A 700 W microwave oven is connected to a 120 V outlet. How much current flows ?
A. 8.4×10^4 A B. 5.8 A C. 0.17 A D. 0.049 A

Problems.

8. A 12 V battery is connected to a 100 Ω resistor.
   A. What is the current in the resistor ?
   B. How many electrons flow through the wire in 10 min ?

9. A 2500 W heater is connected to a 120 V outlet for 10 h.
   A. How much heat is produced in this time ?
   B. What is the current ?
   C. What is the resistance of the heater ?

10. A 4000 W air conditioner is connected to a 220 V source and is operated for 80% of the time for 30 days. If electricity costs $0.10 per kWh, how much does it cost to operate the air conditioner for a month ?

CHAPTER 17 Basic Electric Circuits

Chapter Objectives

Upon completion of the unit on electric circuits, students should be able to :

1. write and apply expressions for the combinations of resistors in parallel, in series, and for series-parallel combinations in order to find the total resistance, total current, total voltage or the current, potential difference , or power for an individual resistor.
2. write and apply Kirchhoff's rules for multiloop circuits in order to determine the current(s) in a circuit, or to find the potential difference or voltage drop across a resistor or group of elements.
3. write and apply expressions for charging or discharging RC circuit for the voltage across the resistor of capacitor of the current in the resistor of charge on the capacitor at any time **t**.
4. sketch graphs for the current and the voltage for a resistor or for charge and voltage as functions of time for RC circuits charging or discharging.
5. find the power and / or energy for a resistor and a capacitor, or the potential difference in a combination of resistors, in a multiloop circuit, or for an RC circuit.
6. draw schematic diagrams for the combinations of resistors, batteries, and capacitors.
7. draw circuits containing properly connected voltmeters and ammeters, and calculate and / or explain the effect of the meter on the circuit.
8. explain the purpose of electrical safety devices.

Chapter Summary

*When resistors are connected in **series** in a circuit, the current is the same through all the resistors, and the equivalent resistance R_s is given by $R_s = R_1 + R_2 + R_3 + \dots$. Combining resistors in series gives the maximum value for the resistance in any combination of resistors.

*When resistors are connected in **parallel** in a circuit, the voltage drop across each resistor is the same, and the reciprocal of the equivalent resistance R_p is given by $1/R_p = 1/R_1 + 1/R_2 + 1/R_3 + \dots$. The equivalent resistance of resistors in parallel is always less than that of the smallest resistor. Combining resistors in parallel gives the minimum resistance for any possible combination.

*A general procedure for analyzing single-battery series-parallel circuits is to find the voltage drops across and currents through the various resistors is :

1. Starting with the resistor combinations farthest from the battery, find the series and parallel equivalent resistances.
2. Reduce the circuit until there is a single loop with one total equivalent resistance.
3. Find the current delivered to the circuit by the battery using Ohm's law.
4. Using this current, expand the circuit by reversing your reduction steps, finding the current and voltage drop for the resistors in each step.

*In a multiloop circuit, a **junction** is a point at which three or more connecting wires are joined together. A circuit path between two junctions is called a **branch** and may contain one or more circuit elements.

***Kirchhoff's rules** are used in analyzing multiloop circuits :

First rule (or junction theorem) : The algebraic sum of the currents at any junction is zero. (conservation of charge)
Second rule (or loop theorem): The algebraic sum of the potential difference around a closed circuit loop is zero. (conservation of energy)

*Sign convention voltage changes across circuit elements in going around a loop :

+V when a battery is transversed toward its positive (+) terminal.
-V when a battery is transversed toward its negative (-) terminal.
-IR when a resistance is transversed in the direction of the assigned branch current.
+IR when a resistance is transversed in the direction opposite to the assigned branch current.

*The general steps in applying Kirchoff's rules to a circuit are :

1. Assign and label a current and current direction for each branch in the circuit. This is done most conveniently at junctions.
2. Indicate the loops and the directions in which the loops are to be transversed. Every branch must be in at least one loop.
3. Applying the junction theorem, write the current equations for each junction that gives a different equation.
4. Transverse the loops applying the loop theorem using the adopted sign convention to determine the signs of the voltage drops.

*Current flows in a series **RC circuit** when the circuit is closed until the capacitor is fully charged and the circuit opened. The charging and discharging times depend on the **time constant** τ, where $\tau = RC$. In one time constant, a capacitor charges to

63% of its maximum value; in two time constants, to over 86% percent, etc. In discharging, a capacitor loses 63% of its total charge in one time constant and by the end of two time constants the capacitor has lost more than 86% of its total charge.

*An dc **ammeter** is a low-resistance instrument that is placed in series with a circuit element to measure the current through it. So large currents can be measured, a "shunt" resistance is placed in parallel with the coil resistance of the galvanometers.

* A dc **voltmeter** is a high-resistance instrument that is placed in parallel with a circuit element to measure the voltage drop across it. A large "multiplier" resistance is placed in series with the coil resistance of the galvanometry so large voltages can be measured.

***Fuses** and **circuit breakers** are safety devices that open or break a circuit when the current exceeds a preset value.

* A **three-prong plug** uses a dedicated grounding wire to ground objects that may become conductors and dangerous. A **polarized plug** identifies the ground side of the line for use as a grounding safety feature.

*The first precaution to take for personal safety is to avoid coming into contact with an electrical conductor that might cause a potential difference to exist across one's body or part of it.

Important Terms and Relationships

17.1 Resistances in Series, Parallel, and Series - Parallel Combinations

resistances in series

equivalent resistance in series : $R_s = R_1 + R_2 + R_3$

$I_t = I_1 = I_2 = I_3$

$V_t = V_1 + V_2 + V_3$

equivalent resistance in parallel : $1/R_p = 1/R_1 + 1/R_2 + 1/R_3$

$I_t = I_1 + I_2 + I_3$

$V_t = V_1 + V_2 + V_3$

17.2 Multiloop Circuits and Kirchoff's Rules

junction
branch

Kirchoff's first rule (junction theorem) : $\Sigma I_i = 0$
Kirchoff's second rule (loop theorem) : $\Sigma V_i = 0$

17.3 RC Circuits

RC circuit
time constant
charging voltage for an RC circuit : $V = V_o (1 - e^{-t/RC})$
discharging voltage for an RC circuit : $V = V_o e^{-t/RC}$
current (charging and discharging) in an RC circuit : $I = I_o e^{-t/RC}$

17.4 Ammeters and Voltmeters

ammeter (dc)
ammeter current : $I_g = IR_s / (r + R_s)$
voltmeter (dc) : $I_g = V / (r + R_s)$

17.5 Household Circuits and Electrical Safety

fuse
circuit breaker
grounded plug
polarized plug

Additional Solved Problems

17.1 Resistances in Series, Parallel, and Series - Parallel Combinations

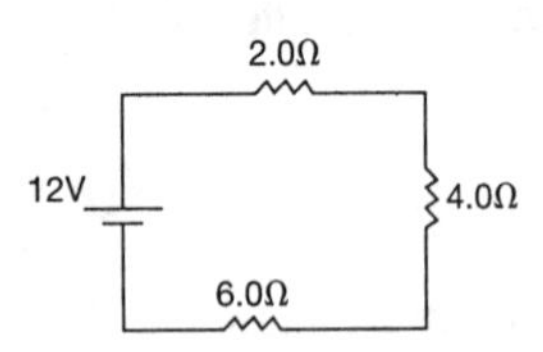

Fig. 17.1

Example 1 Three resistors are connected to a 12-V source as shown in Fig. 17.1.

A. What is the equivalent resistance ?
B. What is the current in the 4.0 Ω resistor ?
C. What is the potential difference across the 6.0 Ω resistor ?

A. $R_s = R_1 + R_2 + R_3$; $R_s = 2.0\ \Omega + 4.0\ \Omega + 6.0\ \Omega = 12\ \Omega$
B. $V_t = I_t R_t$; $12\ V = I_t = (12\ \Omega)$; $I_t = 1.0\ A$
$I_t = I_1 = I_2 = I_3 = 1.0\ A$
C. $V = IR$; $V_{6\ \Omega} = (1.0\ A)(6.0\ \Omega) = 6.0\ V$

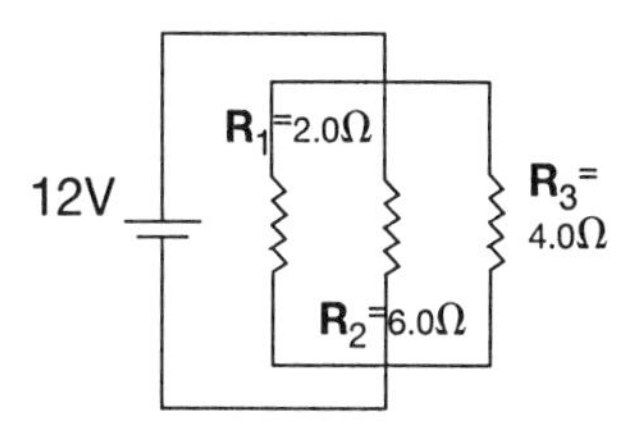

Fig. 17.2

Example 2 Three resistors are connected as shown in Fig. 17.2.
A. What is the equivalent resistance ?
B. What is the voltage across the 6.0 Ω resistor ?
C. What is the current in 4.0 Ω resistor ?

A. $1/R_p = 1/R_1 + 1/R_2 + 1/R_3$
$1/R_p = 1/(2.0\ \Omega) + 1/(6.0\ \Omega) + 1/(4.0\ \Omega)$; $R_p = 1.1\ \Omega$
B. $V_t = V_1 = V_2 = V_3 = 12\ V$
C. $V = IR$; $12\ V = I_{4.0\ \Omega}(4.0\ \Omega)$; $I_{4.0\ \Omega} = 3.0\ A$

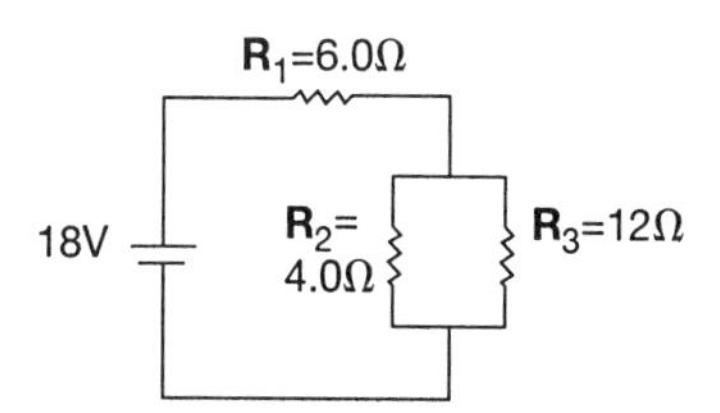

Fig. 17.3

Example 3 A 18 V battery is connected to a circuit as shown in Fig. 17.3.
A. Find the equivalent resistance ?
B. What is the total current delivered by the battery?
C. What is the current in the 4.0 Ω resistor ?
D. What is the power expended in the 6.0 Ω resistor ?

A. First find the equivalent resistance in parallel :

$1/R_p = 1/(4.0\ \Omega) + 1/(12\ \Omega)$; $R_p = 3.0\ \Omega$

$R_s = R_p + R_1 = 3.0\ \Omega + 6.0\ \Omega = 9.0\ \Omega$

B. $V_t = I_t R_t$; $18\ V = I_t (9.0\ \Omega)$; $I_t = 2.0\ A$

C. First find the voltage across the parallel combinations of resistors :

$V_p = (2.0\ A)(3.0\ \Omega) = 6.0\ V$

$6.0\ V = I(4.0\ \Omega)$; $I = 1.5\ A$

D. $P = V^2/R$; $P = 12\ V^2/6.0\ \Omega = 24\ W$

Example 4 A 12-V, 60-W light bulb and a 120-V, 100-W light bulb are connected in (a) parallel and in (b) series. Which bulb burns the brighter in each situation and what is the power rating of the bulb ?

given : $P_1 = 60\ W$; $V_1 = 120\ V$; $P_2 = 100\ W$; $V_2 = 120\ V$

first find the resistance of each bulb :

$P = V^2/R$; $60\ W = (120\ V)^2/R_1$; $R_1 = 240\ \Omega$

$100\ W = (120\ V)^2/R_2$; $R_2 = 144\ \Omega$

(a) in parallel, the voltage is constant and the power ratings are the same as they are individually.

(b) in series : $R = R_1 + R_2 = 240\ \Omega + 144\ \Omega = 384\ \Omega$

$V = IR$; $120\ V = I(384\ \Omega)$; $I = 0.31\ A$

$P_1 = (0.31\ A)^2(240\ \Omega) = 23\ W$

$P_2 = (0.31\ A)^2(144\ \Omega) = 14\ W$

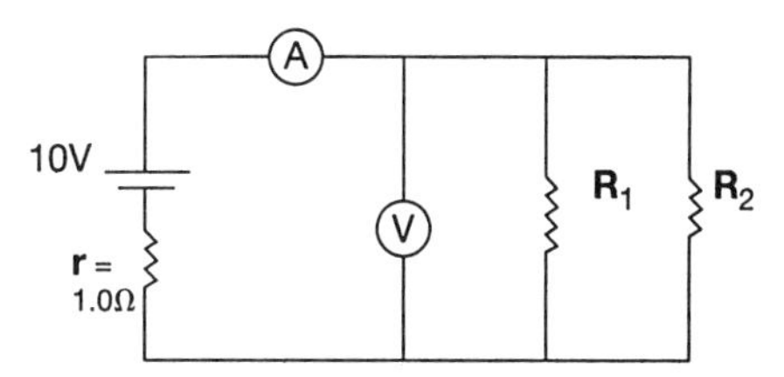

Fig. 17.4

Example 5 In the circuit in Fig. 17.4, the current through the 10 V source is 2.0 A. The voltage source has an internal resistance of 1.0 Ω. The power rating of R_2 is 4.0 W.

A. What is the reading of the voltmeter ?
B. What is the resistance of R_2 ?
C. What is the resistance of R_1 ?

A. first find the verminal voltage : emf - $I_t r$ = terminal voltage

$$10 \text{ V} - (2.0 \text{ A})(1.0 \ \Omega) = 8.0 \text{ V}$$

the terminal voltage is the same as the reading of the voltmeter.

B. $P = V^2 / R$; $4.0 \text{ W} = (8.0 \text{ V})^2 / R_2$; $R_2 = 16 \ \Omega$

C. to find R_1, first find the current in I_2 ; $V = IR$; $8.0 \text{ V} = I_2(16 \ \Omega)$; $I_2 = 0.50 \text{ A}$

$I_t = I_1 + I_2$; $2.0 \text{ A} = I_1 + 0.50 \text{ A}$; $I_1 = 1.5 \text{ A}$

$V = IR$; $8.0 \text{ V} = (1.5 \text{ A}) R_1$; $R_1 = 5.3 \ \Omega$

17.2 Multiloop Circuits and Kirchhoff's Rules

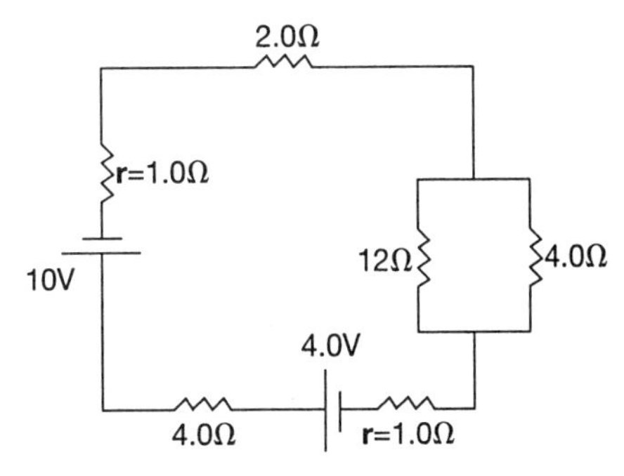

Fig. 17.5

Example 1 In the circuit in Fig. 17.5, each cell has an internal resistance of 1.0 Ω. Find the following for the circuit.

A. the current in the 2.0 Ω resistor.
B. the current in the 12 Ω resistor.

C. terminal voltage for the
(1) 10 V source.
(2) 4.0 V source

A. first use Kirchoff's rules to find I ; start at lower right corner and move clockwise :

$0 = -I(1.0\ \Omega) + 4.0\text{ V} - I(4.0\ \Omega) - I(1.0\ \Omega) - 10\text{ V} - I(2.0\ \Omega) - I(3.0\ \Omega)$

$6.0 = -11\ I$; $I = -0.55$ A negative current means current is counterclockwise.

B. $V_p = IR$; $V_p = (0.55\text{ A})(3.0\ \Omega) = 1.7\text{ V}$

(3.0 Ω is equivalent resisitance in parallel)

now for 12 Ω resistor : $1.7 = I(12\ \Omega)$; $I = 0.14$ A

C. Terminal voltage = emf - Ir

(1) terminal voltage = 10 V - (0.55 A)(1.0 Ω) = 9.5 V

(2) terminal voltage = 4.0 V + (0.55 A)(1.0 Ω) = 4.6 V (charging)

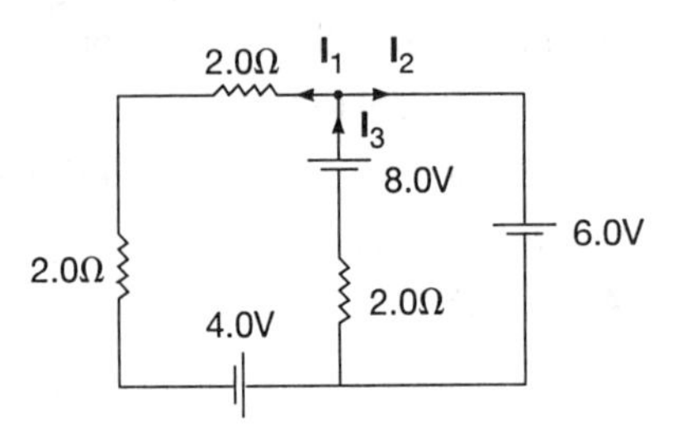

Fig. 17.6

Example 2 In the circuit above, find the current I_1, I_2, and I_3.

start at top in the center :

$I_3 = I_1 + I_2$

$0 = -I_1R - I_1R + 4 - I_3R + 8$

$2I_1R + I_3R = 12\text{ V}$

$4I_1 + (1.0\text{ A})(2.0\ \Omega) = 12$

$I_1 = 2.5$ A

$0 = -6 - I_3R + 8\text{ V}$

$I_3 = 1.0$ A

subsitute back into the first equation : $1.0\text{ A} = 2.5\text{ A} + I_2$; $I_2 = -1.5$ A

17.3 RC Circuits

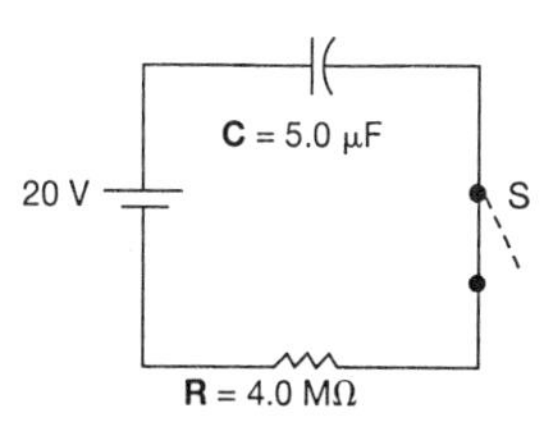

Fig. 17.7

Example 1 For the circuit shown in Fig. 17.7, the switch is closed when the capacitor is initially uncharged.

A. Immediately after the switch is closed, find
   (1) the current in the 4.0 MΩ resistor.
   (2) the potential difference across the capacitor.

B. What is the current in the circuit after 20 s ?

C. After the switch has been closed for a long period of time, find
   (1) the current in the 4.0 MΩ resistor.
   (2) the charge stored in the capacitor.
   (3) the energy stored in the capacitor.

A. right after the switch is closed the capacitor acts similar to a wire ; therefore the voltage across the 4.0 MΩ resistor is 20 V

(1) $V = IR$; $20\ V = I\,(4.0 \times 10^{6}\ \Omega)$; $I = 5.0 \times 10^{-6}\ A$

(2) $V_C = 0$ since $C = Q / V$

B. $I = I_o\,(\ e^{-t/RC})$

$I = (5.0 \times 10^{-6}\ A)\ e^{-20/20}$

$I = 1.8 \times 10^{-6}\ A$

C. after a long period of time, there is no current flowing and the potential across the resistor is zero.

(1) $I = 0$

(2) $C = Q / V$; $5.0 \times 10^{-6}\ F = Q / (20\ V)$; $Q = 1.0 \times 10^{-4}\ C$

(3) $U = (1/2)CV^2$; $U = (1/2)(5.0 \times 10^{-6}\ F)(20\ V)^2 = 1.0 \times 10^{-3}\ J$

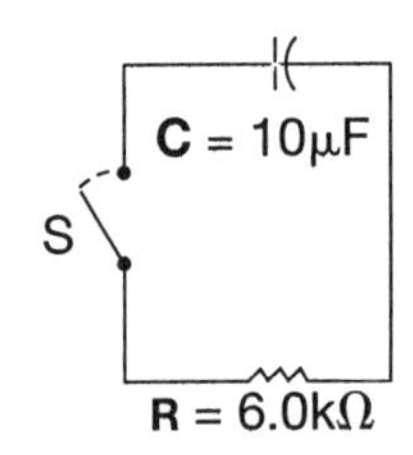

Fig 17.8

Example 2 A 10 μF capacitor stores 2.0×10^{-3} J of energy before the switch in is an RC circuit closed (Fig. 17.8).

A. Immediately after the switch is closed, find
   (1) the current in the 6.0 kΩ resistor.
   (2) the rate energy is being supplied by the capacitor.

B. After the switch has been closed for a long period of time, find
   (1) the current in the 6.0 kΩ resistor.
   (2) total energy expended by the resistor.

A. (1) $U = (1/2)CV^2$; 2.0×10^{-3} J = $(1/2)(10 \times 10^{-6})V^2$; V = 20 V
   V = IR ; 20 V = I $(6.0 \times 10^3\ \Omega)$; I = 3.3×10^{-3} A
   (2) P = IV ; P = $(3.3 \times 10^{-3}$ A)(20 V) = 6.7×10^{-2} W

B. (1) 0
   (2) all of the energy stored in the capacitor goes to the resistor : 2.0×10^{-3} J

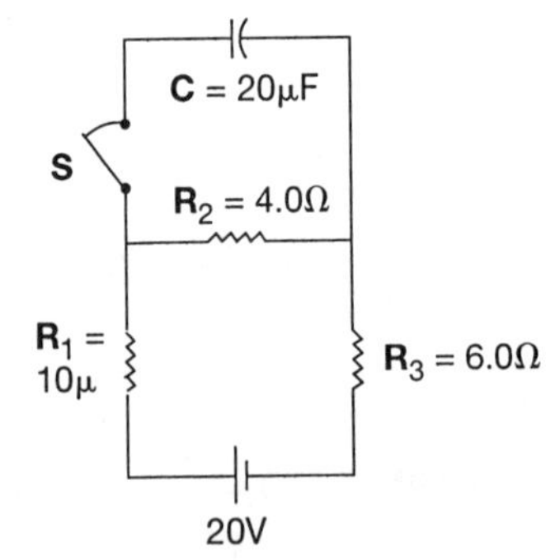

Fig. 17.9

Example 3 A. In the circuit shown in Fig. 17.9, the switch is initally open.
   (1) Find the current in the 6.0 Ω resistor.
   (2) Find the potential difference across the 4.0 Ω resistor.

B. The switch is closed and the circuit reaches steady state values. What is the energy stored in the 2.0 μF capacitor ?

A. (1) $R_s = R_1 + R_2 + R_3 = 10\ \Omega + 4.0\ \Omega + 6.0\ \Omega = 20\ \Omega$

$V_t = I_t R_t$; $20\ V = I_t (20\ \Omega)$; $I_t = 1.0\ A$

(2) $V = IR$; $V = (1.0\ A)(4.0\ V) = 4.0\ V$

B. after a long time no current can flow through the capacitor and all current must again flow through the original path ; the potential across the capacitor equals the potential across the 4.0 Ω resistor.

$C = Q / V$

$20\ \mu F = Q / (4.0\ V)$; $Q = 80\ \mu C$

17.4 Ammeters and Voltmeters

Example A galvonometer with a full-scale sensitivity of 1.0 mA has a coil of resistance of 200 Ω. It is used as an ammeter with full scale reading of 1.5 A. What is the shunt resistance ?

$I_g = IR_s / (r + R_s)$

$(1.0 \times 10^{-3}\ A) = (1.5\ A)(R_s) / (200\ \Omega + R_s)$

$(1.0 \times 10^{-3})\ R_s + 0.200 = (1.5)\ R_s$; $R_s = 0.133\ \Omega$

Solutions to paired problems and other selected problems

4. In a parallel combination, the voltage is constant. Since V = IR, the only situation when the current is the same in each resistor is when the resistors are all equal.

11. $R_{total} = R_{eq} + R$; $7.0\ \Omega = R_{eq} + 4.0\ \Omega$; $R_{eq} = 3.0\ \Omega$;
$1/R_{eq} = 1/6.0\ \Omega + 1/R$; $R = 6.0\ \Omega$

14. $R = R_1 + R_2 + R_3$; $R = 2.0\ \Omega + 4.0\ \Omega + 6.0\ \Omega = 12\ \Omega$

(a) $V_t = I_t R_t$; $12\ V = I_t (120\ \Omega)$; $I_t = 1.0\ A$

(b) the current in a series combination is constant : $I_t = I_1 = I_2 = I_3 = 1.0\ A$

(c) $P = I^2 R$

$P_{2\ \Omega} = (1.0\ A)^2 (2.0\ \Omega) = 2.0$; $P_{4\ \Omega} = (1.0\ A)^2 (4.0\ \Omega) = 4.0\ W$

$P_{6\,\Omega} = (1.0\text{ A})^2(6.0\ \Omega) = 6.0\text{ W}$

(d) $P_{total} = 2.0\text{ W} + 4.0\text{ W} + 6.0\text{ W} = 12\text{ W}$

OR $P = I^2R_t = (1.0\text{ A})^2(12\ \Omega) = 12\text{ W}$

24. (a) $P = IV$; $P_1 = 50\text{ W}$, $P_2 = 100\text{ W}$, $P_3 = 150\text{ W}$;

$50\text{ W} = (120\text{ V})(I_1)$; $I_1 = 0.42\text{ A}$

$100\text{ W} = (120\text{ V})(I_2)$; $I_2 = 0.83\text{ A}$

$150\text{ W} = (120\text{ V})(I_3)$; $I_3 = 1.25\text{ A}$

(b) $P = V^2/R$; $P_1 = 50\text{ W}$, $P_2 = 100\text{ W}$, $P_3 = 150\text{ W}$;

$50\text{ W} = (120\text{ V})^2/R_1$; $R_1 = 288\ \Omega$

$100\text{ W} = (120\text{ V})^2/R_2$; $R_2 = 144\ \Omega$

$150\text{ W} = (120\text{ V})^2/R_3$; $R_3 = 96\ \Omega$

29. first find the heat needed :

$Q = mc\Delta T$; $Q = (0.50\text{ kg})(4190\text{ J/kg-C°})(90\text{ C°}) = 1.89 \times 10^5\text{ J}$

when the power is greatest -- the resistance is least :

$1/R = 1/100\ \Omega + 1/100\ \Omega + 1/100\ \Omega$; $R = 33.3\ \Omega$

now find the power of the heater : $P = V^2/R$; $P = (120\text{ V})^2/33.3\ \Omega = 4.4 \times 10^2\text{ W}$

(a) $P = W/t$ and $W = Q$

$4.4 \times 10^2\text{ W} = (1.89 \times 10^5\text{ J})/t$; $t = 4.3 \times 10^2\text{ s} = 7.2\text{ min}$

(b) the additional heat : $Q = mL = (0.50\text{ kg})(2.26 \times 10^6\text{ J/kg}) = 1.1 \times 10^6\text{ J}$

$4.4 \times 10^2\text{ W} = (1.1 \times 10^6\text{ J})/t_2$; $t_2 = 2.5 \times 10^3\text{ s} = 42\text{ min}$

34. $(1/R_{p1}) = (1/6\ \Omega) + (1/4\ \Omega)$; $R_{p1} = 2.4\ \Omega$;

$R_{s1} = R_{p1} + 2.0\ \Omega = 2.4\ \Omega + 2.0\ \Omega = 4.4\ \Omega$;

$(1/R_{p2}) = (1/R_{s1}) + (1/12\ \Omega)$; $(1/R_{p2}) = (1/4.4\ \Omega) + (1/12\ \Omega)$; $R_{p2} = 3.2\ \Omega$

$(1/R_{p3}) = (1/5\ \Omega) + (1/10\ \Omega)$; $R_{p3} = 3.3\ \Omega$

$R_t = R_{ps} + R_{p2} + 10\ \Omega = 16.5\ \Omega$; $P = V^2/R$; $P = (24\text{ V})^2/16.5\ \Omega = 35\text{ W}$

40. This would give four equations and three unknowns. Any three equations could be used to solve for the unknowns.

46. let I_1 be on left branch in a clockwise direction ;
let I_2 be on the right branch in a clockwise direction ;
let I_3 be in the middle down

equation A : $I_1 = I_2 + I_3$

equation B : $V_1 - I_3R_1 - I_1R_{p1} = 0$

equation C : $-V_2 - I_2R_{p2} + I_3R_1 = 0$

B : $20 - (2/3)I_1 - 5\,i_3 = 0$ $\quad I_1 = (3/2)(20 - 5\,I_3)$

C : $10 + (12/5)\,I_2 - 5\,I_3 = 0$ $\quad I_2 = (5/12)\,(-10 + 5\,I_3)$

substitute into A

$(3/2)(20 - 5\,I_3) = (5/12)(-10 + 5\,I_3) + I_3$

$30 - 7.5\,I_3 = (-50/12) + (25/12)I_3 + I_3$

$I_3 = [\,30 + (50/12)]\,/\,[\,(1 + 7.5 + (25/12)] = 3.23$ A

substituting back into B and C

$I_1 = (3/2)[\,(20 - 5\,(3.2)] = 5.78$ A

$I_2 = (5/12)\,[\,-10 + 5(3.2)] = 2.50$ A

50. (a) in a charging process, the current is greatest just when the switch is closed ; when discharging, the current is greatest just when the switch is opened.

58. (a) when the switch is closed the capacitor has no charge , thus it has no potential and all of the potential must be across the resistor : 12 V
(b) from part (a) the potential across the capacitor is zero.
(c) use Ohm's law to find the current in the circuit : $V = IR$; $12\text{ V} = I\,(6.0\ \Omega)$
$I = 2.0$ A

66. $R_s = I_g r\,/\,(I_{max} - I_g)$; $R_s = [\,(2.0 \times 10^{-3}\text{ A})(100\ \Omega)]\,/\,[3.0\text{ A} - (2.0 \times 10^{-3})]$
$R_s = 0.067\ \Omega$

70. $V = IR$; $R_t = R_a + R$; $I = V\,/\,(R_a + R)$; $I = 6.0\text{ V}\,/\,(10\ \Omega + 0.001\ \Omega)$; $I = 0.59994$ A

<u>Sample Quiz</u>

Multiple Choice. Choose the correct answer.

__ 1. A 2 Ω , a 4 Ω , and a 6 Ω resistor are connected in series to a voltage source. Which resistor expends the most power ?
A. 2 Ω B. 4 Ω C. 6 Ω D. all are the same

__ 2. A 2 Ω, a 4 Ω, and a 6 Ω resistor are connected in parallel to a voltage source. Which resistor expends the most power ?
A. 2 Ω B. 4 Ω C. 6 Ω D. all are the same

__ 3. As more and more resistors are added in series, the total current
A. increases B. decreases C. remains the same

__ 4. A 2 Ω resistor is connected to a 6 V source. A second 2 Ω resistor is connected in parallel with the 6 V source. What is the total current when both resistors are connected ?
A. 1.5 A B. 3.0 A C. 6.0 A D. 12 A

__ 5. As more and more resistors are added in parallel, the total resistance
A. decreases B. increases C. remains the same

A 2.0 μF capacitor is connected in series to a 6.0 MΩ resistor. The capacitor is initially uncharged and the combination is connected to a 20 V source.

__ 6. What is the time constant ?
A. 12 s B. 6.0 s C. 3.0 s D. 0.33 s

__ 7. What is the current flowing through the circuit initially and after the circuit has reached a steady state value ?
A. 0 ; 3.3 μA B. 0 ; 2.0 μA C. 3.3 μA ; 0 D. 2.0 μA ; 0

Problems

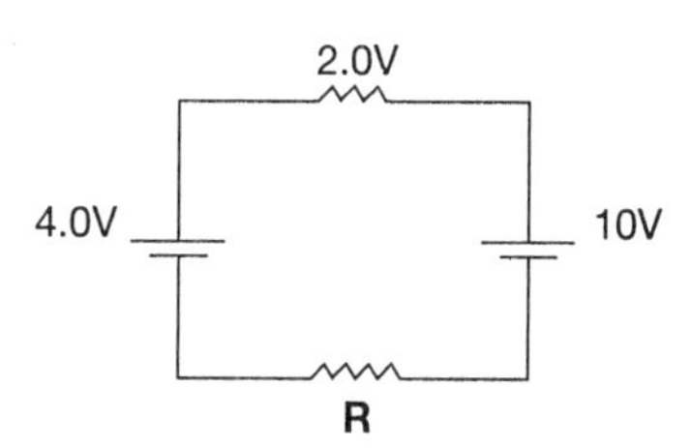

Fig. 17.10

8. In the circuit shown in Fig. 17.10, the current in the 2.0 Ω resistor is 2.0 A.
   A. Find the the value for R.
   B. Find the rate energy is being expended in the 2.0 Ω resistor.
   C. Find the rate energy is being produced in the 10 V cell.

9. A 6.0 Ω resistor and a 12 Ω resistor are connected in parallel and this combination is connected in series with a 2.0 Ω resistor. When a potential difference of 12 V is applied across the combination,
   A. what is the total current ?
   B. what is the current in the 12 Ω resistor ?
   C. what is the potential difference across the 6.0 Ω resistor ?

10. A 2.0 MΩ resistor is connected in series with a 4.0 μF capacitor. This combination is connected to a 40 V source.
   A. What is the current in the circuit immediately after the switche closed ?
   B. What is the current in the circuit after 8.0 s ?
   C. How much energy is stored on the capacitor when it is fully charged ?

CHAPTER 18 Magnetism

Chapter Objectives

Upon completion of the unit on magnetism, students should be able to :

1. describe and apply the law of poles.
2. describe the concept of magnetic dipoles.
3. explain the concept of a magnetic field and sketch magnetic field lines around a single magnet and two magnets.
4. determine the force on a moving charge in a magnetic field and, by using a right-hand rule, find the direction of the force on a moving charge.
5. describe the path of a moving charge in a magnetic field.
6. write expressions for, and apply a right-hand rule to find the magnitude and the direction of the magnetic field around a long, straight wire ; at the center of a circular loop; and for a soleniod.
7. explain magnetic properties using the domain theory for magnetism.
8. state the significance of the Curie point.
9. write and apply the expression for the torque on a current-carrying wire in a magnetic field.
10. understand magnetic properties as they apply to the galvanometer, the dc motor, the cathode ray tube, and the mass spectrometer.
11. describe properties of the Earth's magnetic field.

Chapter Summary

* Magnetism and electricity (electromagnetism) are manifestations of a single fundamental force or interaction, the electromagnetic force.

*Magnets have two different poles or "centers" of force, which are designated as north and south poles. By the **law of poles**, like magnetic poles repel and unlike magnetic poles attract. An isolated, single magnetic pole, or monopole, is postulated but scientists have not verified the existence of the monopole experimentally.

*The direction of the magnetic field vector at any location can be defined as the direction in which the north pole of a compass needle would point. Hence the magnetic field lines are directed away from a magnetic north pole and toward a magnetic south pole.

*The magnitude of the magnetic field can be defined in terms of the magnetic force on a moving electric charge. For a charge q traveling with a velocity v perpendicular to a uniform magnetic field B, the force the charge experiences is F = qvB. Hence, B = F/qv is the force per "moving charge", so to speak, and has the units of N/C-n/s which is given the name of **tesla** (T). This is equal to the older common unit of weber- m^2.

*The direction of the force on a moving charge in a magnetic field is given by a right-hand rule: when the **v** vector is turned or crossed into the **B** vector by the fingers of the right hand, the thumb points in the direction of the force.

*Magnetic fields are produced by electric currents.

* Around a long, straight current-carrying wire, the magnetic field lines are concentric circles, with the field vector tangent to the circular field line at any point on the circle. The direction of the **B** is given by a right-hand rule: when the current-carrying wire is grasped with the right hand with the extended thumb pointing in the direction of the conventional current, the curled fingers indicate the circular sense of the magnetic field.

***Ferromagnetic materials** are easily magnetized. This results from the unpairing of atomic electron spins and the coupling of groups of these atoms into **magnetic domains**. In an external magnetic field, the domains parallel to the field grow at the expense of other domains and the orientation of some domains may become more aligned with the field.

*Common ferromagnetic materials are iron,nickel, and cobalt. Iron is commonly used in the cores of electromagnetics. This type of iron is termed "soft" iron because the domains become unaligned and the iron unmagnetized when the external field is removed. "Hard" iron retains some magnetism after a field is removed, and this type of iron is used to make a permanent magnets by heating the iron above the **Curie temperature** or **Curie point** and cooling in a strong magnetic field. Above the Curie temperature, the domain coupling disappears and a material loses its ferromagnetism.

*A current-carrying wires in a magnetic field experiences a force ($F = ILB \sin\theta$). The direction of the force on the wire is given by a right-hand rule: when the fingers of the right hand are placed in the direction of the conventional current and turned into the B vector, the thumb points in the direction of the force on the wire. The force on a pivoted, current-carrying loop of wire gives rise to a torque ($\tau = IAB \sin\theta$).

*Applications of electromagnetic interactions include the galvanometer, the dc motor, the cathode ray tube, and the mass spectrometer.

*The Earth's magnetic field resembles that which would be produced by a large interior bar magnet (with the magnet's south pole toward the Earth's north-geopraphic compass pole). However, this is not possible because the Earth's interior temperature is above the Curie point of ferromagnetic materials. Scientists associate the Earth's magnetic field with motions (electric currents) in its liquid outer core.

*The magnetic north pole and the geographic north pole do not coincide (nor does the south poles), and the magnetic poles "wander" or move about. The deviation between the direction a compass points and true (geographical) north is expressed in terms of magnetic declination, i.e., the angle between the direction of the magnetic north and geographic north. There is evidence that the Earth's magnetic poles have reversed somewhat periodically over geologic time.

*Charged particles from the Sun and cosmic rays are trapped in the Earth's magnetic field, and regions or concentrations of these charged particles are called Van Allen belts. It is believed that the recombination of ionized air molecules and electrons that have been ionized by particles from the lower belt give rise to the Aurora Borealis (Northern Lights) and the Aurora Australis (Southern Lights.)

Important Terms and Relationships

18.1 Magnets and Magnetic Poles

law of poles
magnetic field

18.2 Electromagnetism and the Source of Magnetic Fields

tesla (T)
magnitude of the magnetic field on a moving charge : $F = qvB \sin \theta$
right - hand for determining the direction of the magnetic force on a moving charge.
magnetic field around a long, straight wire : $B = \mu_o I / 2\pi d$
magnetic field at the center of a circular current-carrying wire : $B = \mu_o I / 2r$
magnetic field in a carrying soleniod : $B = \mu_o n I$

18.3 Magnetic Materials

ferromagnetic materials
magnetic domains
magnetic permeability : $\mu = K_m / \mu_o$
Curie point

18.4 Magnetic Forces on Current-Carrying Wires

force on a straight current-carrying wire : $F = ILB \sin\theta$
right hand rule for determing force on current-carrying wire
torque on a current-carrying loop : $\tau = IAB \sin\theta$

18.5 Applications of Electromagnetism

galvonometer
restoring spring torque in a glavonometer : $\tau_s = k\,\phi$
pointer deflections in a galvonometer : $\phi = NIAB / k$
cathode ray tube (CRT)
mass spectrometer (velocity selector) : $v = E / B$ or $v = V / (Bd)$
particle mass (mass spec) : $m = (qdB_1 B_2 / V)R$

18.6 The Earth's Magnetic Field

Additional Solved Problems

18.2 Electromagnetism and the Source of Magnetic Fields

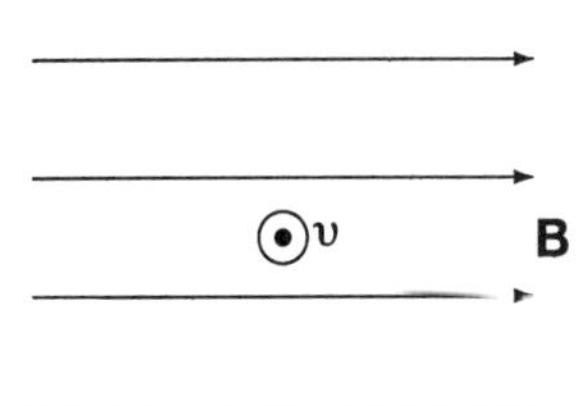

Fig. 18.1

Example 1 A proton has a speed of 1.0×10^6 m/s as it enters a uniform magnetic field of 5.0×10^{-4}T as shown in Fig. 18.1..

A. Sketch the path traveled by the proton.
B. Find the magnetic force on the proton.
C. Find the radius of the circular path traveled by the proton.
D. How long does it take the proton to make a complete circle ?

given : $m = 1.67 \times 10^{-27}$ kg ; $q = 1.6 \times 10^{-19}$ C ; $v = 1.0 \times 10^6$ m/s ;
$B = 5.0 \times 10^{-4}$ T

A. the charge would move in a counterclockwise circle in the plane of the page (righthand rule)

B. $F = qvB$; $F = (1.6 \times 10^{-19}\ C)(1.0 \times 10^6\ m/s)(5.0 \times 10^{-4}\ T) = 8.0 \times 10^{-17}$ N

C. $F = mv^2 / r$

$8.0 \times 10^{-17}\ N = (1.67 \times 10^{-27}\ kg)(1.0 \times 10^6\ m/s)^2 / r$; $r = 21$ m

D. $x = vt$; $2\pi\ (21\ m) = (1.0 \times 10^6\ m/s)\ t$; $t = 1.3 \times 10^{-4}$ s

Example 2 An electron has a speed of 3.0×10^7 m/s perpendicular to a uniform magnetic field. The electron moves in a circle radius 0.20 m in the direction shown on the diagram above. What is the the magnitude of the magnetic field ?

given : $m = 9.1 \times 10^{-31}$ kg ; $q = 1.6 \times 10^{-19}$ C ; $v = 3.0 \times 10^7$ m/s ; $r = 0.20$ m

first find the centripetal force :

$F_c = mv^2 / r = (9.1 \times 10^{-31})(3.0 \times 10^7)^2 / (0.20) = 4.1 \times 10^{-15}$ N now find the magnetic field : $F = qvB \sin 90°$; $4.1 \times 10^{-15} = (1.6 \times 10^{-19})(3.0 \times 10^{7})B$

$B = 8.5 \times 10^{-4}$ T

Example 3 Find the magentic field 2.0 m from a long straight wire which carries a current of 40 A.

given : $d = 2.0$ m ; $I = 40$ A ; $\mu_o = 4\pi \times 10^{-7}$ T-m / A

$B = \mu_o I / 2\pi d$; $B = (4\pi \times 10^{-7}$ T-m / A) (40 A) / [2π (2.0 m)] $= 4.0 \times 10^{-6}$ T

Example 4 What is the magnitude of the magnetic field at the center of a loop of wire which carries a current of 10 A and radius 10 cm ?

given : $I = 10$ A ; $r = 0.10$ m ; $\mu_o = 4\pi \times 10^{-7}$ T-m/A

$B = \mu_o I / 2r$; $B = (4\pi \times 10^{-7}$ T-m/A)(10 A) / [2 (0.10 m)] $= 6.3 \times 10^{-5}$ T

Example 5 A soleniod has 1000 turns per m carries a current of 20 A. What is the magnitude of the magnetic field inside the soleniod ?

given : $\mu_o = 4\pi \times 10^{-7}$ T-m/A ; $N = 1000$ m^{-1} ; $I = 20$ A

$B = \mu_o IN$; $B = (4\pi \times 10^{-7}$ T-m/A)(20 A) (1000 m^{-1}) $= 2.5 \times 10^{-2}$ T

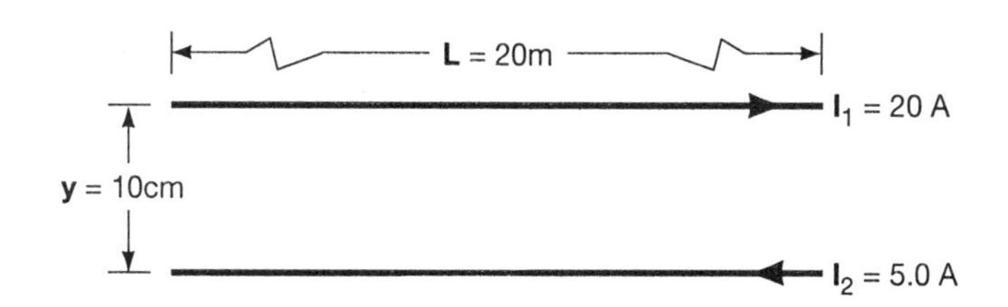

Fig. 18.2

Example 6 Two long straight wire carry currents of 20 A and 5.0 A as shown above. The wires are separated by a distance of 10 cm.

A. What is the magnetic field midway between the two wires ?

B. At what point is the magnetic field zero ?

given : $I_1 = 20$ A ; $I_2 = 5.0$ A ; $y = 10$ cm ; $L = 20$ m

A. B_1 (from top wire) : $B_1 = \mu_o I_1 / 2\pi d$

$B_1 = [(4\pi \times 10^{-7} \text{ T-m/A})(20 \text{ A})] / [2\pi (0.050 \text{ m})]$

$B_1 = 8.0 \times 10^{-5}$ T directed into the page

B_2 (lower wire) : $B_2 = [(4\pi \times 10^{-7} \text{ T-m/A})(5.0 \text{ A})] / [2\pi (0.050 \text{ m})]$

$B_2 = 2.0 \times 10^{-5}$ T directed into the page

$B = B_1 + B_2 = 1.00 \times 10^{-4}$ T directed into the page

B. for the field to be zero, the B fields must be in opposite directions ; it cannot be between the two wires ; it must be below the 5.0 A wire :

$B_1 = B_2$
$I_1 / d_1 = I_2 / d_2$
$20 / (10 + x) = 5.0 / x$
$20 x = 50 + 5.0 x$
$15 x = 50$; $x = 3.3$ cm below the lower wire

18.3 Magnetic Materials

Example A soleniod has 1000 turns per m carries a current of 20 A. What is the magnitude of the magnetic field inside the soleniod if the core has a relative permeability of 1500 ?

given : $N = 1000 \text{ m}^{-1}$; $I = 20$ A ; $K = 1500$; $\mu_o = 4\pi \times 10^{-7}$ T-m/A

$\mu = K \mu_o = (1500)(4 \pi \times 10^{-7} \text{ T-m/A}) = 1.88 \times 10^{-3}$ T-m/A

$B = \mu I N = (1.88 \times 10^{-3} \text{ T-m/A})(20 \text{ A})(1000 \text{ m}^{-1}) = 38$ T

18.4 Magnetic Forces on Current-Carrying Wires

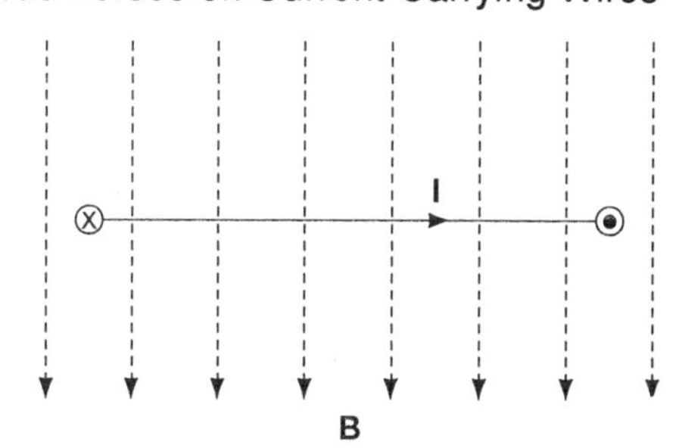

Fig. 18.3

Example 1 A 10.0 m length of wire carries a current of 20.0 A in a uniform magnetic field of 4.0×10^{-3} T as shown in Fig. 18.3. What is the magnetic force on the wire ? How could the wire be rotated so the magnetic force is zero ?

given : I = 20.0 A ; L = 10.0 m ; B = 4.0×10^{-3} T

$F = ILB = (20.0\text{ A})(10.0\text{ m})(4.0 \times 10^{-3}\text{ T}) = 0.80\text{ N}$ directed into the page

If the current in the wire were in the same direction or opposite direction of the magnetic field, the magnetic force would be zero.

Example 2 Two long straight parallel wires carry current of 10 A in the opposite directions. The wires are 3.0 m long and are separated by a distance of 15 cm. What is the force between the two wires ? Is the force attractive or repulsive ?

given : $I_1 = I_2 = 10$ A ; L = 30 m ; d = 0.15 m

$F = \mu_o I_1 I_2 L / 2\pi d$

$F = [\,(4\pi \times 10^{-7}\text{ T-m/A})(10\text{ A})(10\text{ A})(3.0\text{ m})] / [2\,\pi\,(0.15\text{ m})] = 4.0 \times 10^{-4}\text{ N}$

the force is repulsive from right hand rules.

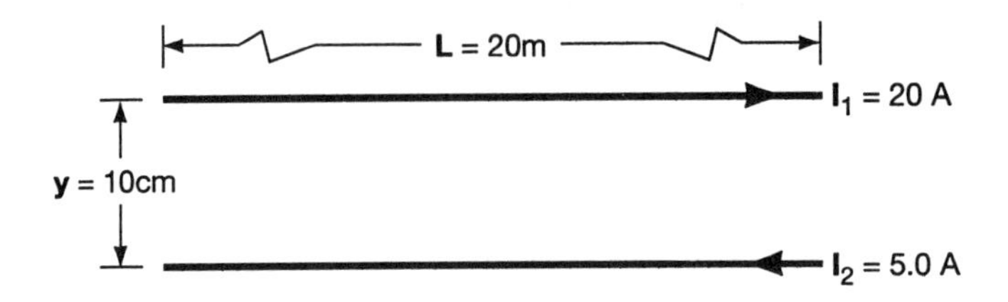

Fig. 18.4

Example 3 What is the force per unit length on the top wire in the Fig. 18.4 ?

given : I_1 = 20 A ; I_2 = 5.0 A ; L = 20 m ; d = 0.10 m

$F / L = \mu_o I_1 I_2 / 2\pi d$

$F / L = [4\pi \times 10^{-7}$ T-m/A)(20 A)(5.0 A)] / [2π (0.10 m)] = 2.0×10^{-4} N / m

18.5 Applications of Electromagnetism

Example 1 An electron passes through crossed electric and magnetic fields. The electric field has a magnitude of 3.0×10^3 V/m and the magnetic field has a magnitude of 4.0×10^{-3} T. What is the speed of the electron ?

given : E = 3.0×10^3 V/m ; B = 4.0×10^{-3} T

v = E / B (since Eq = qvB) ; v = (3.0×10^3 V/m) / (4.0×10^{-3} T) = 7.5×10^5 m/s

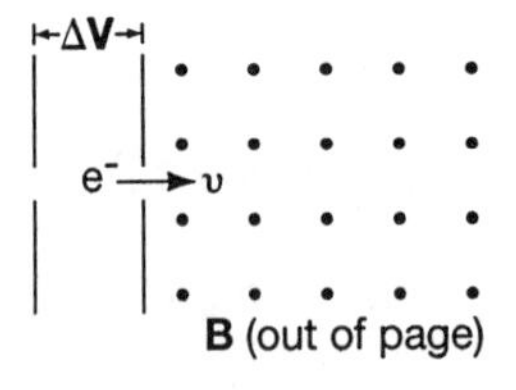

Fig. 18.5

Example 2 An electron is accelerated through a potential to 1000 V and then enters a magnetic field whose magnetic field is 4.0×10^{-3} T as shown in Fig. 18.5.

A. Sketch the path traveled by the electron.
B. Calclate the radius of the path of the electron.

An electric field is now added to straighten the path of the charge.

C. What is the magnitude of the electric field.

D. What is the direction of the electric field.

given : $m = 9.1 \times 10^{-31}$ kg ; $q = 1.6 \times 10^{-19}$ C ; V = 1000 V ; $B = 4.0 \times 10^{-3}$ T

A. the electron would move in a semicircle which would be in a counterclockwise direction.

B. First find the electron's speed ; $V = K / q$; $V = (1/2)mv^2 / q$

$1000 \text{ V} = (1/2)(9.1 \times 10^{-31} \text{ kg}) v^2 / (1.6 \times 10^{-19} \text{ C})$; $v = 1.9 \times 10^7$ m/s

Now work with magnetic forces :

$mv^2 / r = qvB$

$(9.1 \times 10^{-31})(1.9 \times 10^7 \text{ m/s})^2 / r = (1.6 \times 10^{-19} \text{ C})(1.9 \times 10^7 \text{ m/s})(4.0 \times 10^{-3} \text{ T})$

$r = 2.7 \times 10^{-2}$ m

C. $v = E / B$

$1.9 \times 10^7 \text{ m/s} = E / (4.0 \times 10^{-3} \text{ T})$; $E = 7.6 \times 10^4$ N/C

D. The magnetic force is toward the top of the page -- therfore the electric force must be toward the bottom of the page. Since the charge is negative the electric field is opposite direction of the electric force -- thus E is toward the top of the page.

Example 3 In a mass spectrometer, a singly charged ion has a velocity with a magnitude of 1.0×10^6 m/s and is in a magnetic field of 200 mT. The radius of the circular path is 2.0 cm.

A. What is the mass of the ion ?

B. What is the kinetic energy of the ion ?

C. How much work is done on the ion while it is in the magnetic field ?

given : $v = 1.0 \times 10^6$ m/s ; $B = 200 \times 10^{-3}$ T ; $q = 1.6 \times 10^{-19}$ C ; r = 0.020 m

A. $mv^2 / r = qvB$

$m (1.0 \times 10^6 \text{ m/s})^2 / (0.020 \text{ m}) = (1.6 \times 10^{-19} \text{ C})(1.0 \times 10^6 \text{ m/s})(0.200 \text{ T})$

$m = 6.4 \times 10^{-28}$ kg

B. $K = (1/2) mv^2$; $K = (1/2)(6.4 \times 10^{-28} \text{ kg})(1.0 \times 10^6 \text{ m/s})^2 = 3.2 \times 10^{-16}$ J

C. A magnetic force cannot do work (since F is perpendicular to v)

Solutions to paired problems and selected problems

6. the magnetic field would be directed into the page

14. (a) $F = qvB \sin \theta$; $F = (1.6 \times 10^{-19}\ C)(3.0 \times 10^{5}\ m/s)(\sin 37°) = 1.4 \times 10^{-14}\ N$
    $F = ma$; $1.4 \times 10^{-14}\ B = (1.67 \times 10^{-27}\ kg)\ a$; $a = 8.6 \times 10^{12}\ m/s^2$
    (b) the force has the same magnitude but it would be in opposite direction.

18. (a) (1) the charge in negatively charged
    (2) the particle has no charge
    (3) the particle is postively charged
    (b) q/m ratio greater for (3) than for (1)

21. for each wire : $B = \mu_o I / 2\pi d$
    $B = (4\pi \times 10^{-7}\ T\text{-}m/A)\ (8.0\ A) / [\ 2\pi(0.25\ m)] = 6.4 \times 10^{-6}\ T$
    (a) $B = B_1 - B_2$; the fields are in opposite directions and thus cancel ; $B = 0$
    (b) $B = B_1 + B_2 = (6.4 \times 10^{-6}\ T) + (6.4 \times 10^{-6}\ T) = 1.28 \times 10^{-5}\ T$

26. ${d_1}^2 = (12\ cm)^2 + (9\ cm)^2$; $d_1 = 15\ cm$
    $B_1 = \mu_o I / 2\pi d = (4\pi \times 10^{-7}\ T\text{-}m/A)(8.0\ A) / [\ 2\pi(0.15\ m)]$; $B_1 = 1.2 \times 10^{-5}\ T$
    $\tan \theta = 12 / 9$; $\theta = 53°$ relative to the x-axis (origin at i_2, x axis toward i_1)
    $B_2 = (4\pi \times 10^{-7}\ T\text{-}m/A)\ (2.0\ A) / [\ 2\pi\ (9.0 \times 10^{-2}\ m)]$
    $B_2 = 4.4 \times 10^{-6}\ T$ in the x-direction
    $B_x = B_2 + B_1 \cos 53° = (4.4 \times 10^{-6}\ T) + (1.1 \times 10^{-5}\ T) \cos 53° = 11.0 \times 10^{-6}\ T$
    $B_y = B_1 \sin 53° = (1.1 \times 10^{-5}\ T)(\sin 53°) = 8.8 \times 10^{-6}\ T$
    $B^2 = {B_x}^2 + {B_y}^2$; $B = 1.4 \times 10^{-5}\ T$; $\tan \theta = 8.8 / 11$; $\theta = 39°$

34. $F = mv^2 / r$; $F = qvB$; $v = qBr / m$
    $v = (1.6 \times 10^{-19}\ C)(12 \times 10^{-6}\ T)(60 \times 10^{-3}\ m) / (1.67 \times 10^{-27}\ kg) = 69\ m/s$
    $K = (1/2)mv^2$; $K = (1/2)(1.67 \times 10^{-27}\ kg)(69\ m/s)^2 = 4.0 \times 10^{-24}\ J$

42. $m = IA$

$m = (q/t)(\pi r^2)$

$m = e\pi r^2 / 2\pi r/v$

$m = evr / 2$

60. $F / L = \mu_o I_1 I_2 / 2\pi d$

$F / L = [(4\pi \times 10^{-7}\ \text{T-m/A})(2.0\ \text{A})(4.0\ \text{A})] / [2\pi (0.24\ \text{m})] = 6.7 \times 10^{-6}\ \text{N/m}$

67. (a) $m = IA$; $m = (1.5\ \text{A})(0.20\ \text{m})(0.30\ \text{m}) = 9.0 \times 10^{-2}\ \text{A-m}^2$
(b) with plane of coil parallel to the magnetic field

74. $V = K / q$; $250\ \text{V} = K / (1.6 \times 10^{-19}\ \text{C})$; $K = 4.0 \times 10^{-17}\ \text{J}$

$K = (1/2)mv^2$; $4.0 \times 10^{-17}\ \text{J} = (1/2)(9.1 \times 10^{-31}\ \text{kg})\ v^2$; $v = 9.4 \times 10^6\ \text{m/s}$

$E = F / q$; $E = ma / q$; $(0.10\ \text{V/m}) = (9.1 \times 10^{-31}\ \text{kg})\ a / (1.6 \times 10^{-19}\ \text{C})$

$a = 1.8 \times 10^{10}\ \text{m/s}^2$

working with horizontal motion : $x = v_x t$; $0.12\ \text{m} = (9.4 \times 10^6\ \text{m/s})\ t$

$t = 1.3 \times 10^{-8}\ \text{s}$

$y = v_{oy}t + (1/2)at^2 = 0 + (1/2)(1.8 \times 10^{10}\ \text{m/s}^2)(1.3 \times 10^{-8}\ \text{s})^2 = 1.5 \times 10^{-6}\ \text{m}$

80. $v = E / B$; $v = (1.0 \times 10^3\ \text{V/m}) / (0.10\ \text{T}) = 1.0 \times 10^4\ \text{m/s}$

$mv^2 / r = qvB$; $mv / r = qB$; $m\ (1.0 \times 10^4\ \text{m/s}) = (1.6 \times 10^{-19}\ \text{C})(0.10\ \text{T})$

$m = 1.9 \times 10^{-26}\ \text{kg}$

94. gravitational force $w = mg = (9.1 \times 10^{-31}\ \text{kg})(9.80\ \text{m/s}^2) = 8.9 \times 10^{-30}\ \text{N}$

$F = qvB$; $8.9 \times 10^{-30}\ \text{N} = (1.6 \times 10^{-19}\ \text{C})(v)(10^{-4}\ \text{T})$

$v = 5.6 \times 10^{-7}\ \text{m/s}$

Sample Quiz

Multiple Choice. Choose the correct answer.

__ 1. A single charged atom and a double charged ion (same mass) enter with the same velocity into a uniform magnetic field. Which charge will travel in the larger circular path ?
A. single B. double C. paths are the same

__ 2. A magnetic field in the in +x direction and the velocity of a particle is in the -x direction. What is the direction of the force on the particle ?
A. +y B. - y C. +z D. 0

__ 3. The distance from a long straight wire is doubled. By what factor does the magnetic field change ?
A. 1/4 B. 1/2 C. 2 D. 4

__ 4. Two long straight parallel wires carry equal currents. The force between the wires is **F**. The current in each wire is now doubled. What is the new force between the two wires ?
A. **F** / 4 B. **F** / 2 C. 2**F** D. 4**F**

__ 5. Two magnets are held with their opposite poles facing each other. The net force will be
A. attractive B. repulsive C. attractive or repulsive

__ 6. A proton, electron, and helium nucleus (+2e) pass through a velocity selector with equal velocities. Which of the following best descibes what occurs ?
A. all move in circular paths
B. all move in parabolic paths
C. the deflected path of the proton and helium nucleus is in the opposite direction from the electron.
D. all charges travel in straight lines.

__ 7. A 500 turns/m solenoid has a current of 2.0 A . What is the magnitude of the magnetic field inside the solenoid ?
A. 0 B. 1.3×10^{-3} T C. 796 T D. 6.3×10^{-4} T

Problems

8. A proton is subjected to a potential of 800 V and then moves into a magnetic field of 4.0×10^{-3} T.
   A. Calculate the radius the proton travels in the magnetic field.
   B. Calculate the period of motion of the proton.

9. Two long, straight wires carry currents of 10 A and 20 A in the same direction. The wires are separated by a distance of 20 cm and the wires are 3.0 m long.
   A. What is the magnetic field midway between the wires ?
   B. What is the force on the 10 A wire ?

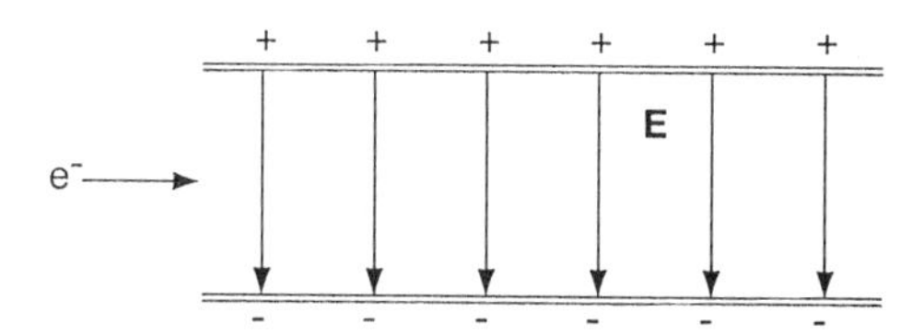

Fig. 18.6

10. In a velocity selector an electron passes through with a speed of 3.0×10^6 m/s in crossed electric and magnetic fields. The electric field is 3.0×10^5 N/C directed in the -y direction as in Fig. 18.6. Find the magnitude and the direction of the magnetic field.

CHAPTER 19 Electromagnetic Induction

Chapter Objectives

Upon completion of the unit on electromagneic induction, students should be able to :

1. state and apply Faraday's law and Lenz's law to find the magnitude and direction of an induced current and to find the magnitude and polarity of the induced emf.
2. describe ways in which a current can be produced.
3. state the purpose of a generator, and write expressions for the emf produced by a generator.
4. explain the parts of a generator and explain how dc and ac current can be produced.
5. explain back emf, list factors which affect the back emf, and explain how a back emf will affect the generated emf.
6. explain how power is transmitted, and the purpose for transformers.
7. write expressions for a step-up transformers and a step-down transformer, and calculate the current and voltage change from a transformer.
8. explain the effect of an eddy current.
9. describe Maxwell's equation in order to describe the symmetry between electric and magnetic fields as they propagate outward in space.
10. classify the types of electromagnetic waves according to increasing frequencies.

Chapter Summary

*Emf's and currents are induced in conducting loops by **electromagnetic induction**, which involves a change in the number of magnetic field lines through the loop.

*A measure of the number of field lines through a particular loop area is given by the **magnetic flux**, $\phi = BA\cos\theta$, where θ is the angle between the B and A vectors. The unit of magnetic flux is the weber- (Wb), or T-m^2.

*The induced emf in a conducting loop depends on the time rate of change of magnetic flux through the loop, i.e., $\varepsilon = -N\Delta\phi/\Delta t$, where N is the number of loops. This relationship is known as **Faraday's law of induction**.

*The minus sign in Faraday's law gives the indication of the induced polarity of the emf, which is found by **Lenz's law**: an induced emf gives rise to a current with a magnetic field that opposes the change in flux producing it.

*Using the flux equation with Faraday's law , we have
$\varepsilon = -N[\ \Delta B/\Delta t)A \cos\theta = + B(\Delta A/\Delta t)\cos\theta + BA\ \Delta(\cos\theta) / \Delta T]$,
and the induced emf can be reproduced in three different ways.

*A **generator** is a device that converts mechanical energy into electrical energy. An ac generator or alternator produces **ac current** for which the polarity of the voltage and direction of the current periodically change. The alternating emf may be represented by $\varepsilon = \varepsilon_0 \sin 2\pi ft$, where f is the rotational frequency of the rotational generator armature. DC current can be produced by a dc generator, although alternating current is usually "rectified" or "converted" it to direct current.

*The rotating armature of a motor in a magnetic field produces a **back emf**, which opposes the line voltage and tends to reduce the current in the armature coils. The back emf depends on the rotational speed of the armature and so helps regulate the motor operation under normal load conditions.

*A **counter torque**, which opposes the armature rotation in a generator, is produced by the interaction of the armature's current-carrying coils in a magnetic field.

* A **transformer** is a device that uses mutual induction to step up or step down ac voltages and currents. This depends upon the relative number of windings on the input primary coil and on the output secondary coil. If the number of turns in the primary is greater than the number of terms in the secondary, the transformer can be classified are being step-down transformer which steps up the voltage and steps down the current. If the number of turns in the primary is greater than the number of terms in the secondary, the transformer is classified as being step-down transformer. In a step-down transformer, the voltage decreases and the current increases. Beside I^2R losses in the transformer windings, another source of energy loss is through **eddy currents** set up in the transformer core.

*In electrical power transmission, the voltage is stepped up and the current is stepped down to reduce the I^2R line losses. Leakage losses through corona discharges limits the magnitude of the voltage increase through corona discharges limits the magnitude of the voltage increase.

*The electric force and the magnetic force are combined into a single force of equations known as **Maxwell's equations**. Basically there is symmetrical relationship which is simply described: a time-varying magnetic field produces a time-varying electric field, and a time-varying electric field produces a time-varying magnetic field.

*The symmetry in Maxwell's equations is important in the analysis of **electromagnetic waves**, which are produced by accelerating electric charges. Electromagnetic waves consists of electric and magnetic fields oscillating at right angles, with both fields perpendicular to the direction of the wave propagation. The interaction of electromagnetic waves with matter gives rise to a force, which per unit area is called **radiation pressure**.

*Electromagnetic waves are classified by ranges of frequencies or wavelengths in a continuous spectrum. The major types or classifications in order of increasing frequency (decreasing wavelength) are power waves, radio (and TV) waves, microwaves, infrared radiation, visible radiation, ultraviolet radiation, X-rays, and gamma rays.

*Infrared radiation is called "heat rays" and is involved in the **greenhouse effect**, which helps regulate the Earth's temperature. Ultraviolet radiation emitted by the Sun is in the large part absorbed by the atmospheric ozone layer. The ultraviolet radiation that reaches the Earth's surface is responsible for sunburns and suntans.

Important Terms and Relationships

19.1 Induced Emf's : Faraday's Law and Lenz's Law

electromagnetic induction
magnetic flux : $\phi = BA \cos \theta$
Faraday's law of induction : $\varepsilon = -N \Delta\phi / \Delta t$
Lenz's law
alternating current

19.2 Generators and Back Emf

ac generator
dc generator
back emf : $\varepsilon_b = V - IR$
counter torque

19.3 Transformers and Power Transmission

transformer
primary coil
secondary coil
step-up transformer
step-down transformer
eddy currents
currents, voltage, and turn ratio : $I_p / I_s = V_s / V_p = N_s / N_p$

19.4 Electromagnetic Waves

electromagnetic waves (radiation)
Maxwell's equations
radiation pressure
power waves
radio and TV waves
microwaves
infrared radiation
visible light
ultraviolet lihgt
X - rays
gamma rays

Additional Solved Problems

19.1 Induced Emf's : Faraday's Law and Lenz's Law

Example 1 A magnet is thrusted into a loop of wire 10 cm in radius. The magnetic field increases from 0 to 3.0 mT in 0.050 s. What is the induced voltage in the loop ?

given : $N = 1.0$; $r = 10$ cm ; $\Delta B = 3.0 \times 10^{-3}$ T ; $t = 0.050$ s

$\varepsilon = -N\Delta BA / t$

$\varepsilon = -(1.0)(+3.0 \times 10^{-3})(\pi)(0.10 \text{ m})^2 / (0.050 \text{ s}) = -1.9 \times 10^{-3}$ V

Example 2 A coil of wire, diameter 20 cm, has 20 turns. The wire is made of copper and has a radius of 1.0 mm. A magnet is pulled from the loop and the magnetic field changes from 0.35 mT to 0 in 0.20 s.

A. Find the induced voltage in the coil.
B. Find the resistance of the coil.
C. Find the current in the wire.
D. Find the charge that flows in the wire.
E. Find the amount of electrical energy produced.

given : $N = 20$; $d = 0.20$ m ; $\rho_{Cu} = 1.7 \times 10^{-8}$ Ω-m ; $r = 1.0 \times 10^{-3}$ m ;

$\Delta B = -0.35 \times 10^{-3}$ T ; $t = 0.20$ s

A. $\varepsilon = -N\Delta BA / t$

$\varepsilon = -(20)\,(-0.35 \times 10^{-3}\ \text{T})(\pi)(0.10\ \text{m})^2 / (0.20\ \text{s}) = 1.1 \times 10^{-3}\ \text{V}$

B. $R = \rho\, L / A = (1.72 \times 10^{-8}\ \Omega\text{-m})(20)(2\pi)(0.10\ \text{m}) / [\ \pi\ (1.0 \times 10^{-3}\ \text{m})^2 = 6.9 \times 10^{-2}\ \Omega$

C. $\varepsilon = IR$; $1.1 \times 10^{-3}\ \text{V} = I\ (6.9 \times 10^{-2}\ \Omega)$; $I = 1.6 \times 10^{-2}$ A

D. $q = It$; $q = (1.6 \times 10^{-2}\ \text{A})(0.20\ \text{s}) = 3.2 \times 10^{-3}$ C

E. $P = IV = (1.1 \times 10^{-3}\ \text{V})(1.6 \times 10^{-2}\ \text{A}) = 1.8 \times 10^{-5}$ W

Example 3 A square coil of wire has 15 turns and an area of 0.40 m^2 is placed perpendicular to a magnetic field of 0.75 T. The coil is flipped so its plane is parallel with the field in 0.050 s. What is the induced voltage in the coil ?

given : $N = 15$; $A = 0.40\ \text{m}^2$; $\Delta B = 0.75$ T ; $t = 0.050$ s

$\varepsilon = -N\Delta BA / t$

$\varepsilon = -(15)(0.75\ \text{T})(0.40\ \text{m}^2) / (0.050\ \text{s}\ ) = -90\ \text{V}$

Example 4 A wire 20 cm in length is moved with a constant velocity of 20 m/s perpendicular to a magnetic field of 0.80 T. What is the induced voltage in the wire ?

given : $r = 0.20$ m ; $v = 20$ m/s ; $B = 0.80$ T

$\varepsilon = BLv$; $\varepsilon = (0.80\ \text{T})(0.20\ \text{m})(20\ \text{m/s}) = 3.2\ \text{V}$

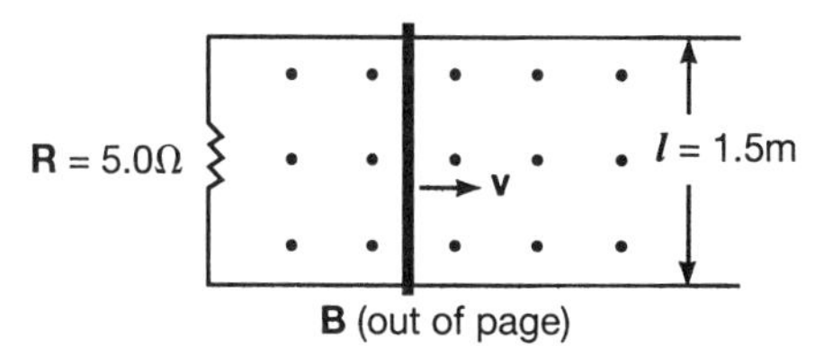

Fig. 19.1

Example 5 As shown in Fig. 19.1, a counductor moves along parallel, frictionless rails connected on one end by a 5.0 Ω resistor. A 0.50 T magnetic field is directed out of the page. The length of the rod is 1.5 m and the velocity of the rod is 3.0 m/s in the +x direction.

given : R = 5.0 Ω ; L = 1.5 m ; v = 3.0 m/s ; B = 0.50 T

A. Calculate the induced voltage in the rod.
B. Calculate the induced current in the rod.
C. Calculate the rate electrical energy is dissipated in the rod.
D. Calculate the rate energy must be provided by the applied force to keep the bar moving with a constant speed.
E. Calculate the force which must be exerted on the bar to keep it moving with a constant speed.
F. Calculate the magnetic force on the bar.

A. $\varepsilon = BLv$; $\varepsilon = (0.5\ T)(1.5\ m)(3.0\ m/s) = 2.25\ V$

B. use Ohm's law : $V = IR$; $2.25\ V = I\ (5.0\ \Omega)$; $I = 0.45\ A$

C. $P = V^2 / R$; $P = (2.25\ V)^2 / 5.0\ \Omega$; $P = 1.0\ W$

D. rate energy is supplied equals rate energy is dissipated in 5.0 Ω resistor.

E. by the law of conservation of mechanical energy $P = 1.0\ W$

F. $F = ILB \sin\theta$; $F = (0.45\ A)(1.5\ m)(0.5\ T) = 0.33\ N$
or $P = Fv$; $1.0\ W = F\ (3.0\ m/s)$; $F = 0.33\ N$; the magnetic force would be in the - x direction.

19.2 Generators and Back Emf

Example 1 A 60 cycle generator has a maximum voltage of 220 V. It is connected to a 20 Ω resistor. At what time after t = 0

A. is the current maximum ?
B. is the current zero ?

given : $f = 60\ Hz$; $V_{max} = 220\ V$; $R = 20\ \Omega$

(a) maximum at (1/4)T and (3/4)T -- 0.013 s and 0.043 s
(b) zero at multiples of T and (1/2)T -- 0.027 s and 0.85 s

Example 2 A dc motor is connected to 24 V power supply. When a external load of 20 Ω is connected, it is found that the back emf is 4.0 V.
A. Find the start-up current.
B. Find the operational current in the motor ?

given : $V = 24\ V$; $\varepsilon_b = 4.0\ V$; $R = 20\ \Omega$

$$\varepsilon_b = V - IR;$$

A. $0 = 24\ V - I\,(20\ \Omega) = 1.2\ A$

B. $4\ V = 24\ V - I\,(20\ \Omega)$; $I = 1.0\ A$

19.3 Transformers and Power Transmission

Example 1 A transformer has an input current of 0.50 A at 120 V. The number of turns on the primary is 10 and the number of turns on the secondary is 120.
A. Find the output current.
B. Find the output voltage.
C. Is this a step-up or a step-down transformer ?

given : $N_p = 10$; $V_p = 120\ V$; $I_p = 0.50\ A$; $N_s = 120$

(a) $N_p/N_s = V_p/V_s$; $(10)/(120) = (120\ V)/V_s$; $V_s = 1.44 \times 10^3\ V$

(b) $N_p / N_s = I_s / I_p$; $(10/120) = I_s / (0.50\ A)$; $I_s = 4.2 \times 10^{-2}\ A$

(c) since the voltage increases and the current decreases , the transfomer is an example of a step-up transformer.

Example 2 A power station produces 6.0 MW of electricity at 20,000 V. The electricity is then sent over power lines which have a resistance of 1.5 Ω.
A. How much electrical energy is lost in the wires ?
B. What voltage is lost over the wires ?

given : $P_{output} = 6.0\ MW$; $V_{output} = 20{,}000\ V$; $R = 1.5\ \Omega$

A. first find the current : $P = IV$; 6.0×10^6 A = I (2.0×10^4 V) ; $I = 3.0 \times 10^2$ A
energy lost due to joule heating : $P = I^2R = (3.0 \times 10^2\ A)^2(1.5\ \Omega) = 1.4 \times 10^5$ W
B. $V = IR$; $V = (3.0 \times 10^2\ A)(1.5\ \Omega) = 4.5 \times 10^2$ V

19.4 Electromagnetic Waves

Example 1 The moon is approximately 3.8×10^5 km from the Earth. Approximately how long does it take the light reflected from the moon to reach the Earth ?

given : $c = 3.0 \times 10^8$ m/s ; $d = 3.8 \times 10^5$ km $= 3.8 \times 10^8$ m

the speed of light is 3.0×10^8 m/s
$d = vt$; 3.8×10^8 m = (3.0×10^8 m/s) t ; t = 1.3 s

Example 2 A certain radio station broadcasts at a frequency of 101.5 MHz. What is the wavelength of the trasmitted signal ?

given : the speed (v) of electromagnetic waves is 3.0×10^8 m/s
$f = 101.5$ MHz $= 1.015 \times 10^8$ Hz

$v = \lambda f$; (3.0×10^8 m/s) = λ (101.5×10^6 Hz) ; $\lambda = 3.0$ m

Solutions to paired problems and other selected problems

8. $\phi = BA \cos\theta =$; $A = (1/2)bh$
$\phi = (0.550\ T)(1/2)(0.40\ m)(0.40\ m)(\sin 60°) = 3.8 \times 10^{-2}$ T-m^2

15. $\varepsilon = -N\, \Delta\phi\, / \Delta t$
9.0 V = - (50) ($\Delta\phi$) / (0.20 s) ; $\phi = -3.6 \times 10^{-2}$ Wb
$\phi = BA$; -3.6×10^{-2} Wb $= \Delta B\, (\pi)(0.10\ m)^2$; $\Delta B = -1.15$ T
$\Delta B = B - B_o$; $-1.15\ T = B - 1.50\ T$; $B = 0.35$ T

18. $\varepsilon = -Blv$

22. first find the potential using Ohm's Law : $V = IR$; $V = (3.0\ A)(0.10\ \Omega)$; $V = 0.30\ V$
$\varepsilon = -N\ \Delta(BA) / \Delta t$; $0.30\ V = -\ (1)\ \Delta B\ (0.060\ m^2) / \Delta t$
$\Delta B / \Delta t = -5.0\ T/s$

30. When the value of the emf is a maximum, the area of the loop is perpendicular to the magnetic field. Since the area of the loop is perpendicular to the plane of the loop, the plane of the loop must be parallel to the field.

37. $\varepsilon = \varepsilon_o \sin 2\pi ft$

(a) $\varepsilon / \varepsilon_o = \sin 2\pi ft = \sin [\ (120\ \pi)(0.50)] = 0$

(b) $\varepsilon / \varepsilon_o = \sin [\ (120\ \pi)(1/360] = 0.87$ or 87%

42. (a) $\varepsilon_b = V - IR$; $10\ V = 12\ V - (I)(0.40\ \Omega)$; $I = 5.0\ A$
(b) $V = IR$; $12\ V = I\ (0.40\ \Omega)$; $I = 30\ A$

47. (a) step down ; $I_p / I_s = N_s / N_p$; $I_p / I_s = 600 / 800 = 3/4$
(b) $I_s = (4/3)I_p = (4/3)(4.0\ A) = 5.3\ A$; $V_s / V_p = 3/4$; $V_3 = (3/4)(120\ V) = 90\ V$

52. (a) $V_s I_s = V_p I_p$; $V_s = (10 / 4)\ (120\ V) = 3.0 \times 10^2\ V$
(b) $N_p = (I_s / I_p)\ N_s$; $N_p = (0.40)(800) = 3.2 \times 10^2$ turns

58. $N_s = (V_s / V_p)\ N_p$; $N_s = (10^4 / 440)\ (132) = 3.0 \times 10^3$ turns

65. X-rays have the greastest frequency and the smallest wavlength of the choices listed ---- (c)

70. $d = vt$; $6.7 \times 10^8 = 240{,}000\ mi / t$; $t = 3.6 \times 10^{-4}\ h = 1.3\ s$
it takes 1.3 s for light to travel each way --- 2.6 s

Chapter 19

Sample Quiz

Multiple Choice. Choose the correct answer.

__ 1. The speed of a moving conductor is doubled. By what factor is the energy dissipated in the counductor changed ?
A. 1/4 B. 1/2 C. 2 D. 4

__ 2. Which of the following types of electromagnetic waves has the greatest wavelength ?
A. red light B. green C. ultraviolet D. blue

__ 3. The number of windings of a loop is doubled while all other parameters are held constant. By what factor is the induced current changed ?
A. 1/4 B. 1/2 C. 2 D. 4

__ 4. Which of the following will not produced an induced current in a wire loop ?
A. squeeze the loop in a magnetic field
B. move a magnetic field into the loop
C. move the loop in a magnetic field
D. hold a magnet stationary in the loop

__ 5. Lenz's law is similar in character to which consrvative law ?
A. angular momentum
B. linear momentum
C. force
D. energy

__ 6. A transfomer has 50 turns on the primary and 200 turns on the secondary. What is the ratio of the voltage in the secondary to the voltage in the primary ?
A. 1/16 B. 1/4 C. 4 D. 16

__ 7. A transfomer has 100 turns on the primary and 10 turns on the secondary. What is the ratio of the current in the primary to the current in the secondary ?
A. 1/100 B. 1/10 C. 10 D. 100

Problems

8. The magnetic field directed perpendicularly through a loop increases from 0.35 mT to 0.80 mT in 50 ms. If the area of the loop is 30 cm^2, what is the induced voltage in the loop ?

9. A conductor with a resistance of 0.20 Ω moves perpendicularly through a 0.40 T magnetic field with a speed of 5.0 m/s. The counductor has a length of 0.20 m.
   A. What is the induced voltage in the conductor ?
   B. What is the induced current in the conductor ?
   C. What is the rate at which electrical energy is dissipated in the conductor ?
   D. What is the rate energy must be provided to pull the conductor through the field ?
   E. What force must be exerted on the counductor to pull it through with constant speed ?

10. A electromagnetic wave has a frequency of 7.0×10^{15} Hz.
   A. How far will the wave travel in 3.0 s ?
   B. What is the wavelength of the wave ?

Chapter 20 AC Current

Chapter Objectives

Upon completion of the unit on ac circuits, students should be able to :

1. describe ac current , and note key points such as rms values and peak values.
2. write expressions and sketch graphs for sinusoidal motion of ac currents and voltages in terms of time.
3. calculate rms and peak values for ac current, voltage, and power.
4. write expressions for and apply the expressions for capacitive reactance and inductive reactance.
5. apply Ohm's law to capacitive and inductive reactance.
6. describe impedance, and sketch a phase diagram for the resistance, capacitive reactance, and inductive reactance in order to find the impedance.
7. write expressions for the impedance and Ohm's law relationship for the circuit.
8. find the phase angle between the voltage and the current, and explain when the circuit is inductive and when the circuit is capacitive.

Chapter Summary

*For the dc relationships of Ohm's law ($V = IR$) and power ($P=IV = I^2R = V^2/R$) to apply to ac currents and voltages, special time-average values must be used. These are $P = I^2_{rms}R$ and $V_{rms} = I_{rms}R$, where rms subscript denotes root-mean-square values.

*The impeding effect of a capacitor to current flow in an ac circuit is expressed in terms of **capacitance reactance** (X_CT), and $X_C = (1/2)\pi fC$, with the unit of ohm (Ω). In Ohm's law form, we have $V = IX_C$, where V and I are rms values.

*The impeding effect of an inductor or coil in an ac circuit is expressed in terms of **inductive reactance** (X_L). and $X_L = 2\pi fL$, where L is the inductance with the unit of henry (H). The reactance has the unit of ohm (Ω), and $V = X_L I$, where V and I are rms values.

*In a purely capacitive ac circuit, the current leads the voltage by 90°. In a purely inductive ac circuit, the voltage leads the current by 90°. A helpful phrase to remember which quantity leads the other is ELI the ICE man, where E represents voltage.

*For an ac series RLC circuit, reactance and resistance are conveniently called **phasors.** The phasor sum is the effective opposition or impedance (Z) to the current flow and $Z = [R^2 + (X_L - X_C)^2]^{1/2}$, where the **phase angle** ϕ is given by $\phi = (X_L - X_C)/R$. If $X_L > X_C$, giving a $(+\phi)$, the circuit is inductive. If $X_C > X_L$, giving a $(-\phi)$, the circuit is capacitive.

*The power dissipated by the resistance in an ac series RLC circuit is $P = IV\cos\phi$, where $\cos\phi = R / Z$ is called the **power factor**. This is a maximum in a purely resistive circuit $(Z = R)$.

*The impedance is a minimum at the **resonance frequency**, $f_o = 1/2\pi\,[LC]^{1/2}$, in which case, $X_C = X_L$. At resonance, the circuit is completely resistive and there is maximum power transfer to (and dissipated in) the circuit.

Important Terms and Relationships

20.1 Resistance in an AC Circuit

peak voltage
peak current

rms or effective current :	$I_{rms} = I_o / 2^{1/2}$
rms or effective voltage :	$V_{rms} = V_o / 2^{1/2}$
instantaneous voltage in an ac circuit :	$V = V_o\, 2\pi ft$
instantaneous current in an ac circuit :	$I = I_o\, 2\pi ft$
effective power in an ac circuit :	$P = I_{rms}^2 R$
Ohm's law for ac :	$V_{rms} = I_{rms}R$

20.2 Capacitive Reactance

capacitive reactance X_C :	$X_C = 1 / 2\pi fC$
Ohm's law :	$V = IX_C$

20.3 Inductive Reactance

inductive reactance X_L :	$X_L = 2\pi fL$
Ohm's law	$V = IX_L$

20.4 Impedance : RLC Circuits

phase diagram	
phasors	
impedance (Z)	$Z^2 = R^2 + (X_L - X_c)^2$
phase angle (ϕ)	$\tan\phi = (X_L - X_c) / R$
inductive circuit	
capacitive circuit	
Ohm's law for RC, RL, and RLC circuits	$V = IZ$
power factor	$\cos\phi = R / Z$

20.5 Circuit Resonance

resonance frequency	$f_o = 1 / [2\pi\ (LC)^{1/2}]$

Additional Solved Problems

20.1 Resistance in an AC Circuit

Example 1 A hair dryer rated at 1500 W is connected to a 120 V outlet.
A. What are the rms and the peak values for the currents through the dryer ?
B. What is the resistance of the dryer ?
C. What is the maximum power which can be produced ?

given : $P_{rms} = 1500$ W ; $V_{rms} = 120$ V

A. $P = IV$; $1500\text{ W} = (I_{rms})\ (120\text{ V})$; $I_{rms} = 12.5$ A

$I_o = 2^{1/2}\ (12.5\text{ A}) = 1.77$ A

B. $P = V^2 / R$; $1500\ W = (120\ V)^2 / R$; $R = 9.60\ \Omega$

C. $P = I_o V_o = (2^{1/2}) I_{rms} (2^{1/2}) V_{rms} = 2\ P_{rms} = 2\ (1500\ W) = 3.00 \times 10^3\ W$

Example 2 The maximum voltage for an ac circuit is 1.5 A when connected to a maximum voltage of 170 V.
A. What is the resistance ?
B. What is the rms voltage ?
C. What is the rms current ?

given : $I_o = 1.5\ A$; $V_o = 170\ V$

A. $V = IR$; $170\ V = (1.5\ A)\ R$; $R = 113\ \Omega$

B. $V_o = V_{rms}\ (2)^{1/2}$; $170\ V = V_{rms}\ (2)^{1/2}$; $V_{rms} = 120\ V$

C. $I_o = I_{rms}\ (2)^{1/2}$; $1.5 = (I_{rms})\ (2^{1/2})$; $I_{rms} = 1.1\ A$

20.2 Capacitive Reactance

Example A 500 μF capacitor is connected to a 220 V , 60 Hz source. What is
(a) the capacitive reactance ?
(b) the current in the circuit ?

given : $C = 500\ \mu F$; $V = 220\ V$; $f = 60\ Hz$

(a) $X_C = 1 / (2\pi f C)$

$X_C = 1 / [\ 2\pi\ (60\ Hz)(500 \times 10^{-6}\ F)] = 5.3\ \Omega$

(b) $V = i\,X_C$; $220\ V = i\ (5.3\ \Omega)$; $i = 41.5\ A$

20.3 Inductive Reactance

Example A 50 mH inductor is connected to a 220 V, 60 Hz source. What is
(a) the inductive reactance
(b) the current in the circuit ?

given : $L = 50 \times 10^{-3}\ H$; $V = 220\ V$; $f = 60\ Hz$

(a) $X_L = 2\pi fL = 2\pi\ (60\ Hz)(50 \times 10^{-3}\ H) = 18.8\ \Omega$

(b) $V = I\,X_L$; $220\ V = I\ (18.8\ \Omega)$; $I = 11.7\ A$

20.4 Impedance : RLC Circuits

Example 1 A series RL circuit has a 20 Ω resistor and a 50 mH inductor. How much current flows in the circuit when connected to a 120 V, 60 Hz source ?

given : $R = 20\ \Omega$; $L = 50 \times 10^{-3}\ H$; $X_L = 18.8\ \Omega$ (from previous section)

$Z^2 = R^2 + X_L{}^2$; $Z^2 = (20\ \Omega)^2 + (18.8\ \Omega)^2$; $Z = 27.4\ \Omega$

$V = IZ$; $120\ V = I\ (27.4\ \Omega)$; $I = 4.4\ A$

Example 2 If the inductor in Example 1 is replaced with a 500 μF capacitor, how much current would then flow in the circuit ?

given : $C = 500\ \mu F$; $X_C = 5.3\ \Omega$; $V = 120\ V$

$Z^2 = R^2 + X_C{}^2$; $Z^2 = (20\ \Omega)^2 + (5.3\ \Omega)^2$; $Z = 20.7\ \Omega$

$V = i\,Z$; $120\ V = i\ (20.7\ \Omega)$; $i = 5.8\ A$

Example 3 If the inductor and the capacitor given in example 1 and 2 are placed in series with the resistor, what is
A. the impedance of the circuit ?
B. the current in the circuit ?
C. the phase angle between the current and the voltage supplied ?

from two previous problems $X_C = 5.3\ \Omega$; $X_L = 18.8\ \Omega$; $R = 20\ \Omega$

A. $Z^2 = R^2 - (X_L - X_C)^2$

$Z^2 = (20\ \Omega)^2 - [\ (18.8\ \Omega) - (5.3\ \Omega)]^2$; $Z = 14.8\ \Omega$

B. $V = IZ$; $120\ V = I\ (14.8\ \Omega)$; $I = 8.1\ A$

C. $\tan \phi = (X_L - X_c) / R$

$\tan \phi = [(18.8 \ \Omega) - (5.3 \ \Omega)] / 20 \ \Omega$

$\phi = 34°$

20.5 Circuit Resonance

Example A series LRC circuit has a 100 Ω resistor, a 500 μF capacitor, and a 50 mH inductor.
A. What is the resonant frequency source with an output of 50 V ?
B. How much current flows in the circuit when it is in resonance ?
C. What is the potential difference across the
i. inductor ?
ii. capacitor ?
iii. resistor ?

given : R = 100 Ω ; L = 50 mH and hence $X_L = 18.8 \ \Omega$;
C = 500 μF and hence $X_C = 5.3 \ \Omega$

A. $f_o = 1 / [2\pi (LC)^{1/2}]$; $f_o = 1 / \{ 2\pi [(50 \times 10^{-3} \text{ H})(500 \times 10^{-6} \text{ F})^{1/2}] = 32 \text{ Hz}$

B. $V = i Z$; $50 \text{ V} = i (100 \ \Omega)$; $i = 0.50 \text{ A}$

i. $V_L = IX_L = (0.50 \text{ A})(18.8 \ \Omega) = 9.4 \text{ V}$

ii. $V_C = IX_C = (0.50 \text{ A})(5.3 \ \Omega) = 2.7 \text{ V}$

iii. $V = IR = (0.50 \text{ A})(100 \ \Omega) = 50 \text{ V}$

Solutions to paired problems and selected problems

2. (c)

10. given : P = 1600 W ; V = 120 V ;
(a) $P = IV$; $1600 \text{ W} = I (120 \text{ V})$; $I = 13 \text{ A}$
(b) $I_o = I_{rms}(2^{1/2})$; (13 a) (1.4) = 18 A
(c) $P = V^2 / R$; $1600 \text{ W} / (120 \text{ V})^2 / R$; $R = 9.0 \ \Omega$

16. given : $I = 8.0 \sin 4\pi t$; $V = 60 \sin 4\pi t$

$I_o = I_{rms}(2)^{1/2}$; $8.0\ A = I_{rms}(2)^{1/2}$; $I_{rms} = 5.7\ A$

$V_o = V_{rms}(2)^{1/2}$; $60 = V_{rms}(2)^{1/2}$; $V_{rms} = 42\ V$

$P = IV = (5.7\ A)(42\ V) = 2.4 \times 10^2\ W$

25. (d)

32. (a) $X_L = 2\pi fL$; $X_L = 2\pi\ (60\ Hz)(0.050\ H) = 19\ \Omega$

(b) $V = IX_L$; $120\ V = I\ (19\ \Omega)$; $I = 6.3\ A$

(c) 90°

39. $X_L = 2\pi fL$; $X_C = 1 / 2\pi fC$

$2\pi fL = 1 / 2\pi fC$

$L = 1 / 4\pi^2 f^2 C$; $L = 1 / [\ 4\pi^2(60\ Hz)^2(10^{-7}\ F) = 0.70\ H$

42. (c)

46. (a) $X_c = 1 / 2\pi fC$; $X_C = 1 / [\ 2\pi(60\ Hz)(25 \times 10^{-6})]$; $X_C = 106\ \Omega$

(b) $Z^2 = R^2 + X_c{}^2$; $Z^2 = 200\ \Omega^2 + 106\ \Omega^2$; $Z = 226\ \Omega$

$V = IZ$; $120\ V = I\ (226\ \Omega)$; $I = 0.53\ A$

56. $f = 1 / 2\pi[LC]^{1/2}$; $980 \times 10^3\ Hz = 1 / 2\pi\ [(2.5 \times 10^{-3})\ C]^{1/2}$; $C = 1.1 \times 10^{-11}\ F$

62. $P = I^2R$; $10.0\ W = i^2(60.0\ \Omega)$; $I = 0.408\ A$

$V = IZ$; $120\ V = (0.408\ A)(Z)$; $Z = 294\ \Omega$

$Z^2 = R^2 + X_L{}^2$; $294\ \Omega^2 = 60\ \Omega^2 + X_L{}^2$; $X_L = 288\ \Omega$

$X_L = 2\pi fL$; $288 = 2\pi(60\ Hz)\ L$; $L = 764\ mH$

74. (a) $X_L = 2\pi fL$; $X_L = 2\pi(60\ Hz)(0.100\ H) = 38\ \Omega$

(b) $Z^2 = R^2 + X_L{}^2$; $Z^2 = 50\ \Omega^2 + 38\ \Omega^2$; $Z = 63\ \Omega$

(c) $V = IZ$; $120\ V = i\ (63\ \Omega)$; $I = 1.9\ A$

(d) zero

Sample Quiz

Multiple Choice. Choose the correct answer.

__ 1. The rms current for a circuit is 1.0 A. What is the peak current ?
A. 0.50 A B. 0.71 A C. 1.4 A D. 2.0 A

__ 2. A circuit has an impedance of 60 Ω and a resistance of 30 Ω. What is the power factor ?
A. 2 B. 1 C. 0.5 D. 0.71

__ 3. An inductor has a inductance of 10 mH. What is the capacitive reactance when connected to a 90 Hz source ?
A. 0.27 Ω B. 0.18 Ω C. 5.7 Ω D. 3.8 Ω

__ 4. The rms voltage for a ac circuit is 190 V and the maximum current is 2.0 A. What is the resistance ?
A. 67 Ω B. 95 Ω C. 134 Ω D. 190 Ω

__ 5. The frequency for a circuit containing an inductor is doubled. By what factor does the inductive reactance change ?
A. 1/4 B. 1/2 C. 2 D. 4

__ 6. The frequency of an output source is doubled for a circuit containig a capacitor. By what factor does the capacitive reactance change ?
A. 1/4 B. 1/2 C. 2 D. 4

__ 7. What is the resonant frequency for a circuit containing a 10 mH inductor and a 100 μF capacitor ?
A. 160 Hz B. 1.6×10^{5} Hz C. 6.3×10^{-6} Hz D. 6.3×10^{-3} Hz

Problems

8. An ac voltage is applied to a 20 Ω resistor, and it dissipates 100 W of power. Find the effective voltage and peak current in resistor.

9. A 2.0 μF capacitor is connected across a 120 V voltage source and a current of 40 mA is measured to flow through the capacitor. What is the capacitive reactance of the circuit ?

10. A resistor, an inductor, and a capacitor have values of 20 Ω , 10 mH, and 200 μF, respectivley. These are connected in series to a 120 V, 60 Hz power supply.
    A. Find the current in each element.
    B. Find the potential difference across each element.

CHAPTER 21 Geometric Optics : Reflection and Refraction of Light

Chapter Objectives

Upon completion of the unit on geometric optics, students should be able to :

1. define the term geometric optics.
2. state and apply the law of reflection.
3. describe regular, specular, and irregular (diffuse) reflection.
4. explain Huygens' principle.
5. describe the property of refraction when the light is bent toward the normal and away from the normal.
6. state and apply Snell's law for the refraction for light.
7. determine the speed of light in various substances when given the index of refraction.
8. explain the critical angle, and the property of internal reflection.
9. apply Snell's law to determine the critical angle.
10. explain some uses of internal reflection.
11. explain why light is dispersed.

Chapter Summary

*A **wave front** is the line or surface defined by adjacent portions of a wave that are in phase.

*A **ray** is a line drawn perpendicular to a series of wave fronts in the direction of propagation. The use of such geometric wave representations is describing optical phenomena is called **geometric optics**.

*The **law of reflection** states: the angle of incidence is equal to the angle of reflection, i.e., $\theta_i = \theta_r$. These angles are measured relative to a line drawn perpendicular or <u>normal</u> to the reflecting surface. The incident ray, the reflected ray, and the normal all lie in the same plane.

***Regular**, or **specular reflection** occurs from smooth surfaces, with the reflected rays being parallel. **Irregular**, or **diffuse reflection** occurs from rough surfaces, with the reflected rays being at different angles.

***Refraction** refers to the "bending" or change in direction of a wave boundary when it passes from one medium to another medium as a result of different wave speeds in the different media.

*Huygens' principle for wave propagation is: every point on an advancing wave front can be considered a source of secondary waves, or wavelets, and the envelope tangent to these wavelets defines a new position of the wave front.

***Snell's law** relates the angle of incidence θ_1 and the angle of refraction θ_2 to the wave speeds in the respective media, $\sin\theta_1 / \sin\theta_2 = v_1/v_2$. The **index of refraction** n is the ratio of the speed of light in a vacuum to the speed of light v in a medium, $n = c/v$, or $n = \lambda/\lambda_m$, where λ is the wavelength of light in a vacuum and λ_m the wavelength in a material. In terms of the index of refraction, Snell's law is $\sin\theta_1/ \sin\theta_2 = n_2 / n_1$.

*At a certain critical angle θ_c, the angle of refraction for a ray going from a medium of greater optical density to a medium of lesser optical density is 90° and the refracted ray is along the media boundary. For incident angles $\theta_1 > \theta_c$, the **internal reflection** of light occurs and the surface acts like a mirror. By Snell's law, the critical angle of a medium is related to the indices of refraction: $\sin\theta_c = n_2 / n_1$ (where $n_1 > n_2$).

For air $\sin\theta_c = 1 / n$. Total internal reflection is the principle of **fiber optics**.

***Dispersion** is the spreading out of light into its component wavelength because a material has different indices of refraction for different wavelengths. A rainbow is produced by refraction, dispersion, and internal reflection within water droplets.

Important Terms and Relationships

21.1 Wave Front and Rays

wave front
plane wave front
ray

21.2 Reflection

angle of incidence
angle of reflection
law of reflection : $\theta_i = \theta_r$
regular (specular) reflection
irregular (diffuse) reflection

21.3 Refraction

refraction
Huygen's principle
angle of refraction
Snell's law $n_1 \sin \theta_1 = n_2 \sin \theta_2$ or $\sin \theta_1 / \sin \theta_2 = v_1 / v_2$
index of refraction $n = c / v = \lambda / \lambda_m$

21.4 Total Internal Reflection and Fiber Optics

critical angle $\sin \theta_c = 1/n$
total internal reflection
fiber optics
fiberscope

21.5 Dispersion

dispersion

Additional Solved Problems

21.2 Reflection

Example 1 A beam of light makes an angle of 40° with respect to a surface. What is the angle of reflection ?

solution :

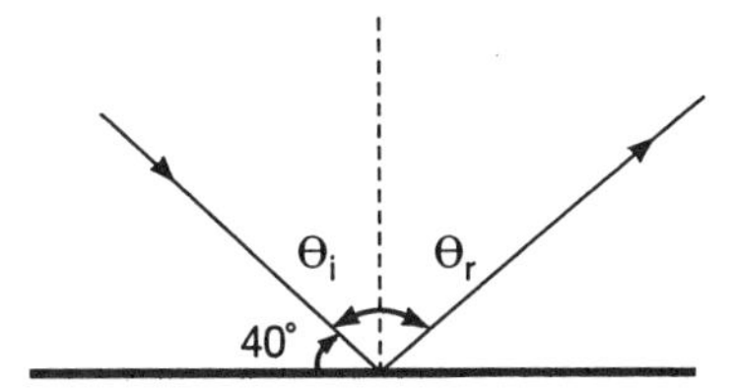

Fig. 21.1

$$\theta_r = \theta_i = 90° - 40° = 50°$$

Example 2 A person is 6 ft 6 in. tall. A person stands 6 ft from a plane mirror. His eyes are 6 in. from the top of his head. Find the minimum height of the mirror if the person is to see his complete image.

From his eyes to the top of his head, since triangles are similar, the height of the top of the mirror from the top of his head should be 3 in. Then sketch a ray of light from his eye to the bottom of his feet. When the normal is sketched, there are more similar triangles and the bottom of the mirror is 3 ft from the floor.

3 ft + 3 in. = 3 ft. 3 in.

21.3 Refraction

Example 1 A certain substance has an index of refraction of 1.8. What is the speed of light in the substance ?

given : $n = 1.8$; $c = 3.0 \times 10^8$ m/s

$n = c / v$; $1.8 = (3.0 \times 10^8 \text{ m/s}) / v$; $v = 1.7 \times 10^8$ m/s

Example 2 The wavelength of a beam of light is 500 nm in air. The beam then enters glass (index of refraction = 1.5).
A. What is the frequency of the light in air ?
B. What is the speed of the light in the glass ?
C. What is the frequency of the light in the glass ?
D. What is the wavelength of the light in the glass ?

given : $n_{air} = 1.0$; n_{glass} 1.5 ; $\lambda_{air} = 500$ nm ; $c = 3.0 \times 10^8$ m/s

(a) $v = \lambda f$; 3.0×10^8 m/s = $(500 \times 10^{-9})(f)$; $f = 6.0 \times 10^{14}$ Hz
(b) $n = c / v$; $1.5 = (3.0 \times 10^8$ m/s$) / v$; $v = 2.0 \times 10^8$ m/s
(c) the frequency of the light stays the same not matter the medium : 6.0×10^{14} Hz
(d) $v = \lambda f$; 2.0×10^8 m/s = λ $(6.0 \times 10^{14}$ Hz) ; $\lambda = 3.3 \times 10^{-7}$ m

Example 3 A beam of light passes from air into water. The angle of incidence is 40°. What is the angle of refraction ?

given : n_1 (air) = 1.0 ; $\theta_1 = 40°$; n_2 (water) = 1.33 ; find θ_2

$n_1 \sin\theta_1 = n_2 \sin\theta_2$

$(1.0)(\sin 40°) = (1.33) \sin\theta_2$; $\theta_2 = 29°$

Example 4 A beam of light passes from a material which has an index of refraction of 2.0 to air.
A. If the angle of incidence is 25°, what is the angle of refraction ?
B. If the angle of incidence is 40°, what is the angle of refraction ?

given : $n_1 = 2.0$; $n_2 = 1.0$

(a) $n_1 \sin\theta_1 = n_2 \sin\theta_2$

$(2.0) \sin 25° = 1.0 (\sin\theta_2)$; $\theta_2 = 58°$

(b) $(2.0) \sin 40° = (1.0)(\sin\theta_2)$; θ_2 = undefined (light is not refracted -- more about this in the next section)

Example 5 A beam of light strikes a piece of glass (n = 1.6) 5.0 cm thick with an angle of incidence of 30°.
A. At what angle does the light enter the glass ?
B. At what angle does the light exit the glass ?
C. What is the distance between the emerging ray relative to the normal ?

given : $n_1 = 1.6$; $d = 5.0$ cm ; $\theta = 30°$

(a) $n_1 \sin \theta_1 = n_2 \sin \theta_2$

$(1.0) \sin 30° = (1.6) \sin \theta_2$; $\theta_2 = 19°$

(b) $n_2 \sin \theta_2 = n_3 \sin \theta_3$

$(1.6)(\sin 19°) = (1.0) \sin \theta_3$; $\theta_3 = 30°$

(c) $\tan 19° = x / 5.0$ cm ; $x = 1.7$ cm

21.4 Total Internal Reflection and Fiber Optics

Example 1 A beam of light passes from water (n = 1.33) into air (n = 1.0). What is the largest possible angle of incidence which will produce a refracted ray ?

$n_1 \sin \theta_1 = n_2 \sin \theta_2$

$1.33 (\sin \theta_1) = 1.0 (\sin 90°)$; $\theta_1 = \theta_c = 49°$

Example 2 A beam of light passes is incident on a piece of glass (n = 1.5) at a 48° angle. Explain what happens to the light as it hits the glass-air surface.

$n_1 \sin \theta_1 = n_2 \sin \theta_2$

$(1.5)(\sin 48°) = (1.0) \sin \theta_2$; θ_2 = undefined ; the light is reflected at a 48° angle with the normal

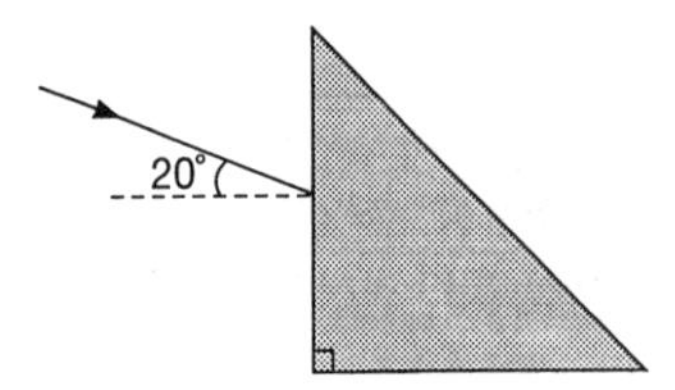

Fig. 21.2

Example 3 A beam of light strikes a 45°-45°-90° prism as shown in Fig.21.2. Trace the path light would travel in the prism. (n for the prism is 1.5)

let θ_2 be the angle of refraction when the light enters the prism :

$n_1 \sin \theta_1 = n_2 \sin \theta_2$; $(1.0) \sin 20° = (1.5) \sin \theta_2$; $\theta_2 = 13°$

the angle the light makes with the surface of the prism is 32° , therefore the new angle of incidence is 58°

$n_1 \sin \theta_1 = n_2 \sin \theta_2$; $1.5 (\sin 58°) = 1.0 (\sin \theta_2)$; θ_2 = undefined - light is reflected off of the surface and now strikes the bottom of the prism :
the angle the light makes with the lower surface is 3° ;

$n_1 \sin \theta_1 = n_2 \sin \theta_2$; $1.5 (\sin 13°) = (1.0) \sin \theta_2$; $\theta_2 = 20°$

22.5 Dispersion

Example A beam of white light strikes a piece of glass at a 30° angle. Red light wavelength 680 nm and blue light wavelength 430 nm emerge after being dispersed. The index of refraction for the red light is 1.55 and the index of refraction for blue light is 1.52.
A. What is the angle of refraction for each color ?
B. Which color of light is refracted the most ?

(a)

blue : $n_1 \sin \theta_1 = n_2 \sin \theta_2$; $(1.0)(\sin 30°) = (1.52) \sin \theta_2$; $\theta_2 = 19.2°$

red : $n_1 \sin \theta_1 = n_2 \sin \theta_2$; $(1.0) (\sin 30°) = (1.55) \sin \theta_2$; $\theta_2 = 18.8°$

(b) by inspection in part (a) the blue light is refracted the most.

Solutions to paired problems and selected problems from the text

4. (d) all of the staements are true

12. (a) 35°
(b) 55°

16. (d)

21. $n = c / v$; $n = (3.0 \times 10^8 \text{ m/s}) / (2.15 \times 10^8 \text{ m/s})$; $n = 1.40$

28. $n_1 \sin \theta_1 = n_2 \sin \theta_2$
$(\sin 30°)(1.0) = (\sin 45°)\, n^2$
$n_2 = 1.4$

34. $v_a n_a = v_b n_b$
$v_a(4/3) = v_b\,(5/4)$
$v_b / v_a = 1.07$

42. $d_{app} / d_{act} = v_{app} / v_{act}$; $n = c / v$
$d_{app} / 3.5 = 1 / 1.33$
$d_{app} = 2.6$ m

46. $n_1 \sin \theta_1 = n_2 \sin \theta_2$
$1.33\,(\sin 40°) = (1.0) \sin \theta_2$; $\theta_2 = 59°$; altitude = 90° - 59° = 31°

51. (b)

54. $\sin \theta_r = \sin 37° / 1.515$; $\theta_r = 23.41°$
$\sin \theta_p = \sin 37° / 1.523$; $\theta_p = 23.28°$
$\Delta\theta = 0.13°$

Sample Quiz

Multiple Choice. Choose the correct answer.

__ 1. Light enters air from water. The beam of light as it enters the air will be
A. greater than or equal to the angle of incidence
B. less than or equal to the angle of incidence
C. greater than the angle of incidence
D. less than the angle of incidence

__ 2. Compared with the speed of light in water the speed of light in air is
A. greater   B. less than   C. equal to   D. always constant

__ 3. A ray of light makes a 42° angle with the surface of a plane mirror. What is the angle of reflection ?
A. 0°   B. 42°   C. 48°   D. 58°

Refers to questions 4 and 5

When light passes from an unknown substance into air the angle of refraction is 90° when the angle of incidence is 37°

__ 5. What is the index of refraction for the unknown substance ?
A. 0.60   B. 0.80   C. 1.25   D. 1.67

__ 6. If the angle of incidence is increased to 42°, what happens to the light ?
A. refracted at an angle between 70° and 90°
B. refracted at an angle greater than 90°
C. reflected at a 48° angle
D. reflected at a 42° angle

__ 7. When light passes through a prism, which color of light is refracted the most ?
A. blue   B. green   C. red   D. yellow

Problems

8. Light passes from air into zircon (n = 1.9) with an angle of incidence of 30°. The wavelength of the light is 500 nm.
   A. What is the speed of light in zircon ?
   B. What is the frequency and the wavelength of the light in zircon ?
   C. What is the angle of refraction ?

9. Light passes from glass (n = 1.5) into water (n = 1.33). Find the angle above which light will be internally reflected.

10. Light enters the rectangular piece of glass (n = 1.8) at an angle incidence of 37° . The thickness of the glass is 10 cm.
    A. What is the angle of refraction when the light enters the glass ?
    B. What is the angle of refraction as the light exits the glass ?
    C. What is the distance the path of light is changed in comparison to its intial path ?

CHAPTER 22 Mirrors and Lenses

Chapter Objectives

Upon completion of the unit on mirrors and lenses, students should be able to :

1. apply similar traiangles to quantitative studies of plane mirrors.
2. distinguish between a real image and a virtual image.
3. write an expression for the magnification of an object.
4. name the two types of spherical mirrors, and locate the focal point, radius of curvature, and vertex for the mirrors.
5. draw parallel rays and radial rays to locate an image for a concave and convex mirror.
6. state and apply the spherical lens equation.
7. name the two types of sperical lenses, and locate the focal point and the optical center on the lenses.
8. draw parallel rays and radial rays to locate an image for a converging and a diverging lens.
9. state and apply the thin lens equation.
10. explain lens aberrations.
11. state and apply the lens maker's equation.
12. write the expression for, and find the power of a lens in diopters.

Chapter Summary

*Images are classified as being real or virtual. A **real image** is one formed by light rays that converge at and pass through the image position. (A real image can be seen or formed on a screen.) A **virtual image** is one for which the light rays appear to emanate from the image, but does not actually do so. The image of a plane mirror is a good example of a virtual image. (A virtual image cannot be seen or formed on a screen.)

*Image characteristics are (1) either real or virtual, (2) upright or inverted, (3) magnified, the same size , or reduced in size. The lateral magnification (M) is M = image height / object height = h_i / h_o . M may be greater than, less than one, or equal to one.

*Spherical mirrors can be concave or convex, depending on which side of the spherical surface is a mirror. A concave mirror is called a **converging mirror** and a convex mirror is called a **diverging mirror**. These terms refer to the reflection of rays parallel to a mirror's optical axis. For spherical mirrors, the focal point is one-half of the radius of curvature.

*The images formed by spherical mirrors can be determined through ray diagrams. A **parallel ray** is a ray parallel to the optical axis reflected through the focal point (as are all rays near and parallel to the axis). A **chief** or **radial ray** is a ray through the center of curvature C along the radial line from C. Image characteristics can also be determined analytically using the spherical mirror equation and the magnification equation with an appropriate sign convention.

*A spherical bioconvex lens is called a **converging lens**, and spherical biconcave lens is called a **diverging lens.** Images formed by these lenses can be determined through ray diagrams. A **parallel ray** is a ray parallel to the lens axis that after refraction (a) passes through the focal point on the opposite or image side of a converging lens, or (b) appears to diverge from the object side focal length of a diverging lens. A **chief ray** is a ray that passes through the center of the lens undeviated. Image characteristics can also be determined analytically using the thin lens equation and the magnification equation with an appropriate sign convention. These equations have the same form as those for spherical mirrors. However, the focal length is <u>not</u> one-half the radius of curvature for spherical lens as is the case for spherical mirrors.

*Lenses may be used in combinations, in which case the image of the first lens may be considered to be the object for the second lens. The total magnification is given by the product of the individual magnifications.

*Rays passing through real lens may not exactly follow the particular rays in a ray diagram because of aberrations. **Spherical aberration** results when parallel pays passing through different regions of a lens do not come together at a common focus. (Spherical aberration also occurs for spherical mirrors for rays far from the optical axis.) **Chromatic aberration** results because the index of refraction of a lens is not the same for all wavelengths or is dispersive. **Astigmatism** results when a cone of light from an off-axis source falls on a lens surface and forms an elliptical illuminated area, giving rise to two images.

*The **lens maker's equation** is a general equation for determining the focal length of a spherical lens with sides having different radii of curvature.

*Lens <u>power</u> is expressed in terms of **diopters**, which is the reciprocal of the focal length of a lens in meters.

Important Terms and Relationships

22.1 Plane Mirrors

plane mirror
virtual image
real image
lateral magnification

22.2 Spherical Mirrors

spherical mirror
concave (converging) mirror
convex (diverging) mirror
center of curvature
radius of curvature
focal point
focal length : $f = R / 2$
parallel ray (mirror)
focal ray (mirror)
spherical mirror equation : $1/f = 1/d_o + 1/d_i$
magnification factor (spherical mirror)
spherical aberration (mirror)

22.3 Lenses

lens
converging (bioconvex) lens
diverging (bioconcave) lens
parallel ray (lens)
chief ray (lens)
focal ray (lens)
thin lens equation $1/f = 1/d_o + 1/d_i$
magnification factor $M = - d_i / d_o$
total magnification with a two-lens system $M_t = |M_1| \times |M_2|$

22.4 Lens Aberrations

spherical aberration (lens)
astigmatism
chromatic aberration

22.5 The Len's Maker's Equation

lens maker's equation : $1/f = (n-1)[\,(1/R_1) - (1/R_2)]$

diopters
lens power : $P = 1/f$ (f in m)

Additional Solved Problems

22.1 Plane Mirrors

Example A 10-cm tall object is placed 100 cm in front of a plane mirror.
A. What is the location of the image ?
B. What is the height of the image ?
C. If the image real or virtual ?

(a) for a plane mirror : $d_o = d_i$ = 100 cm behind the mirror

(b) for a plane mirror : $h_o = h_i$ = 10 cm

(c) the image for a plane mirror is vitrual

Sperical Mirrors

Example 1 A 2.0-cm tall object is placed 10 cm in front of a concave mirror, whose radius of curvature is 30 cm.
A. What is the focal length of the mirror ?
B. Where is the image located ?
C. What is the height of the image ?

given : h_o = 2.0 cm ; d_o = 10 cm ; r = 30 cm

(a) r = 2f ; 30 cm = 2 f ; f = 15 cm

(b) $1/f = 1/d_o + 1/d_i$

$1/(15\text{ cm}) = 1/(10\text{ cm}) + 1/d_i$; d_i = -30 cm

(c) $h_i / h_o = d_i / d_o$

$h_i / (2.0\text{ cm}) = (-30\text{ cm}) / (10\text{ cm})$; h_i = -6.0 cm

Example 2 A 5.0-cm tall object is placed at the following locations in front of a concave mirror whose focal length is 10 cm. What is the location and the height of the object when the object is located

A. 30 cm in front of the mirror ?
B. 20 cm in front of the mirror ?
C. 15 cm in front of the mirror ?
D. 10 cm in front of the mirror ?
E. 5 cm in front of the mirror ?

given : (for all parts) $h_o = 5.0$ cm ; $f = 10$ cm

(a) $1/f = 1/d_o + 1/d_i$

$1/(10 \text{ cm}) = 1/(30 \text{ cm}) + 1/d_i$; $d_i = 15$ cm

$h_i / h_o = d_i / d_o$; $h_i / (5.0 \text{ cm}) = (15 \text{ cm}) / (30 \text{ cm})$; $h_i = 2.5$ cm

(b) $1/f = 1/d_o + 1/d_i$

$1/(10 \text{ cm}) = 1/(20 \text{ cm}) + 1/d_i$; $d_i = 20$ cm

$h_i / h_o = d_i / d_o$; $h_i / (5.0 \text{ cm}) = (20 \text{ cm}) / (20 \text{ cm})$; $h_i = 5.0$ cm

(c) $1/f = 1/d_o + 1/d_i$

$1/(10 \text{ cm}) = 1/(15 \text{ cm}) + 1/d_i$; $d_i = 30$ cm

$h_i / h_o = d_i / d_o$; $h_i / (5.0 \text{ cm}) = (30 \text{ cm}) / (15 \text{ cm})$; $h_i = 10$ cm

real image

(d) $1/f = 1/d_o + 1/d_i$

$1/(10 \text{ cm}) = 1/(10 \text{ cm}) + 1/d_i$; d_i is undefined ; no image

(e) $1/f = 1/d_o + 1/d_i$

$1/(10 \text{ cm}) = 1/(5.0 \text{ cm}) + 1/d_i$; $d_i = -10$ cm

$h_i / h_o = d_i / d_o$; $h_i / (5.0 \text{ cm}) = (-10 \text{ cm}) / (5.0 \text{ cm})$; $h_i = -10$ cm

virtual image

Example 3 An object is placed 20 cm in front of a concave mirror. It produces a virtual image which is 30 cm from the mirror.

A. What is the focal length of the mirror ?
B. What is the radius of curvature of the mirror ?

given : $d_o = 20$ cm ; $d_i = -30$ cm (since the image is virtual)

(a) $1/f = 1/d_o + 1/d_i$

$1/f = 1/(20 \text{ cm}) + 1/(-30 \text{ cm})$; $f = 60$ cm

(b) $r = 2f$; $r = 2\,(60 \text{ cm}) = 120$ cm

Example 4 A convex mirror produces a virtual image 10 cm from a mirror. If the object is 20 cm from the mirror, what is the focal length of the mirror ?

given : d_i = -10 cm ; d_o = 20 cm

$1/f = 1/d_o + 1/d_i$

$1/f = 1/(20 \text{ cm}) + 1/(-10 \text{ cm})$; $f = -20$ cm

Example 5 A convex mirror has a focal length of -10 cm. A 10 cm tall object is placed 20 cm in front of the mirror. What is the height and location of the image ?

given : f = -10 cm ; d_o = 20 cm ; h_o = 10 cm

$1/f = 1/d_o + 1/d_i$

$1/(-10 \text{ cm}) = 1/(20 \text{ cm}) + 1/d_i$; $d_i = -6.7$ cm

$h_i / h_o = d_i / d_o$; $h_i / (10 \text{ cm}) = (-6.7 \text{ cm}) / (20 \text{ cm})$; $h_i = -3.35$ cm (virtual image)

22.3 Lenses

Example 1 A 2.0-cm tall object is placed in 10 cm in front of a convex lens , whose focal length is 30 cm.

A. Where is the image located ?

B. What is the height of the image ?

given : h_o = 2.0 cm ; d_o = 10 cm ; f = 30 cm

(a) $1/f = 1/d_o + 1/d_i$

$1/(30 \text{ cm}) = 1/(10 \text{ cm}) + 1/d_i$; $d_i = -15$ cm

(b) $h_i / h_o = d_i / d_o$

$h_i / (2.0 \text{ cm}) = (-15 \text{ cm}) / (10 \text{ cm})$; $h_i = -3.0$ cm

Example 2 A 5.0-cm tall object is placed at the following locations in front of a convex lens whose focal length is 10 cm. What is the location and the height of the object when the object is located

A. 30 cm in front of the lens ?

B. 20 cm in front of the lens ?

C. 15 cm in front of the lens ?

D. 10 cm in front of the lens ?

E. 5 cm in front of the mirror ?

given : (for all parts) $h_o = 5.0$ cm ; $f = 10$ cm

(a) $1/f = 1/d_o + 1/d_i$

$1/(10 \text{ cm}) = 1/(30 \text{ cm}) + 1/d_i$; $d_i = 15$ cm

$h_i / h_o = d_i / d_o$; $h_i / (5.0 \text{ cm}) = (15 \text{ cm}) / (30 \text{ cm})$; $h_i = 2.5$ cm

(b) $1/f = 1/d_o + 1/d_i$

$1/(10 \text{ cm}) = 1/(20 \text{ cm}) + 1/d_i$; $d_i = 20$ cm

$h_i / h_o = d_i / d_o$; $h_i / (5.0 \text{ cm}) = (20 \text{ cm}) / (20 \text{ cm})$; $h_i = 5.0$ cm

(c) $1/f = 1/d_o + 1/d_i$

$1/(10 \text{ cm}) = 1/(15 \text{ cm}) + 1/d_i$; $d_i = 30$ cm

$h_i / h_o = d_i / d_o$; $h_i / (5.0 \text{ cm}) = (30 \text{ cm}) / (15 \text{ cm})$; $h_i = 10$ cm

real image

(d) $1/f = 1/d_o + 1/d_i$

$1/(10 \text{ cm}) = 1/(10 \text{ cm}) + 1/d_i$; d_i is undefined ; no image

(e) $1/f = 1/d_o + 1/d_i$

$1/(10 \text{ cm}) = 1/(5.0 \text{ cm}) + 1/d_i$; $d_i = -10$ cm

$h_i / h_o = d_i / d_o$; $h_i / (5.0 \text{ cm}) = (-10 \text{ cm}) / (5.0 \text{ cm})$; $h_i = -10$ cm

virtual image

Example 3 An object is placed 20 cm in front of a convex lens. It produces a virtual image which is 30 cm from the lens. What is the focal length of the lens ?

given : $d_o = 20$ cm ; $d_i = -30$ cm (since the image is virtual)

$1/f = 1/d_o + 1/d_i$

$1/f = 1/(20 \text{ cm}) + 1/(-30 \text{ cm})$; $f = 60$ cm

Example 4 A concave lens produces a virtual image 10 cm from a lens. If the object is 20 cm from the lens, what is the focal length of the lens ?

given : $d_i = -10$ cm ; $d_o = 20$ cm

$1/f =1/d_o + 1/d_i$

$1/f = 1/(20 \text{ cm}) + 1/(-10 \text{ cm})$; $f = -20$ cm

Example 5 A concave lens has a focal length of -10 cm. A 10-cm tall object is placed 20 cm in front of the lens. What is the height and location of the image ?

given : $f = -10$ cm ; $d_o = 20$ cm ; $h_o = 10$ cm

$1/f = 1/d_o + 1/d_i$

$1/(-10 \text{ cm}) = 1/(20 \text{ cm}) + 1/d_i$; $d_i = -6.7$ cm

$h_i / h_o = d_i / d_o$; $h_i / (10 \text{ cm}) = (-6.7 \text{ cm}) / (20 \text{ cm})$; $h_i = -3.3$ cm (virtual image)

22.5 The Lens Maker's Equation

Example A convex lens made of glass ($n = 1.5$) has a radius of curvature of 40 cm on one side and 30 cm on the other side. What is the radius of curvature on both lens surfaces ?

given : $n = 1.5$; $R_1 = 40$ cm ; $R_2 = 30$ cm

$1/f = (n - 1) [(1/R_1) - (1/R_2)]$

$1/f = (1.5 - 1.0) [(1/40 \text{ cm}) - (1/30 \text{ cm})]$

$f = -240$ cm

Solutions to paired problems and selected problems

2. A plane mirror (b) produces a virtual, upright, unmagnified image.

7. (a) given : $d_o = 2.0$ m

for a plane mirror : $d_o = d_i = 2.0$ m ; behind the mirror

(b) $v_r = (0.50 \text{ m/s}) - (-0.50 \text{ m/s}) = 1.0$ m/s

object jumps at mirror, and image appears to jump in opposite direction.

12. For a vertical mirror, the minumum length is one-half the height of the student.

$L = (1/2)\ (1.5\ m) = 0.75\ m$

For 20° tilt,

$L' = L \cos 20^\circ$
$L' = 0.75\ m\ (\cos 20^\circ) = 0.70\ m$

20. given : $h_o = 3.0$ cm ; $d_o = 20$ cm ; f = 15 cm (since R = 2f)
$1/f = 1/d_o + 1/d_i$
$1 / 15\ cm = 1 / 10\ cm + 1 / d_i$; $d_i = 60$ cm
$M = - d_i / d_o = 60\ cm / 20\ cm = 3.0$
$M = h_i / h_o$; $3.0 = h_i / 3.0\ cm$; $h_i = 9.0$ cm
the image is real and inverted since h_i and d_i are positive

24. (a) $1/f = 1/d_i + 1/d_o$; $(1/15) = (1/40) + (1/d_i)$; $d_i = 24$ cm
$M = -d_i / d_o = -24 / 40 = -0.60$; the image is real and inverted
(b) $(1/15) = (1/30) + (1/d_i)$; $d_i = 30$ cm ; $M = -(30/30) = -1.0$; real and inverted
(c) $(1/15) = (1/20) + (1/d_i)$; $d_i = 60$ cm ; $M = -(60/20) = -3.0$; real and inverted
(d) $(1/15) = (1/15) + (1/d_i)$; $di = \infty$; $M = \infty$
(e) $(1/15) = (1/5.0) + (1/d_i)$; $di = -7.5$ cm ; $M = -(-7.5 / 5.0) = 1.5$
image is virtual and upright

32. given : $h_o = 4.5$ cm ; $d_o = 12$ cm ; $h_i = 9.0$ cm
$M = h_i / h_o$; $M = (9.0\ cm) / (4.5\ cm) = +2.0$
$1/f = 1/d_o + 1/d_i$
$1/f = 1/12\ cm + 1/-24\ cm$
f = 24 cm ; concave mirror with a radius of curvature 2f = 48 cm

36. first work with concave

$1/f = 1/d_o + 1/d_i$

$1/f = 1/d_o + 1 / -1.8\ d_o$; $1/f = 0.45 / d_o$; $f = (2.22)d_o$

now work with convex

$-1/f = 1./d_o + 1d_i$

$-(1/2.22\ d_o) = 1/d_o + 1/d_i$

$-0.45 / d_o = 1/d_o + 1/d_i$; $d_i / d_o = 1 / 1.45$; $d_i / d_o = 0.69$

42. A virtual image will be formed by (d) both (a) and (c).

46. (a) $1/f = 1/d_o + 1/d_i$; $1/22\text{ cm} = 1/15\text{ cm} + 1/d_i$; $d_i = -47\text{ cm}$

$h_i / h_o = d_i / d_o$

$h_i / 4.0\text{ cm} = (47\text{ cm}) / (15\text{ cm})$; $h_i = 12\text{ cm}$

the image is virtual and magnified

(b) $1 / f = 1d_o + 1/d_i$

$1/22\text{ cm} = 1/ 36\text{ cm} + 1/ d_i$; $d_i = 57\text{ cm}$;

$h_i / h_o = d_i / d_o$; $h_i / (4.0\text{ cm}) = (57\text{ cm}) / (36\text{ cm})$; $h_i = 6.3\text{ cm}$

diagram for (a) is the same sketch as in Fig. 22.16d in the text

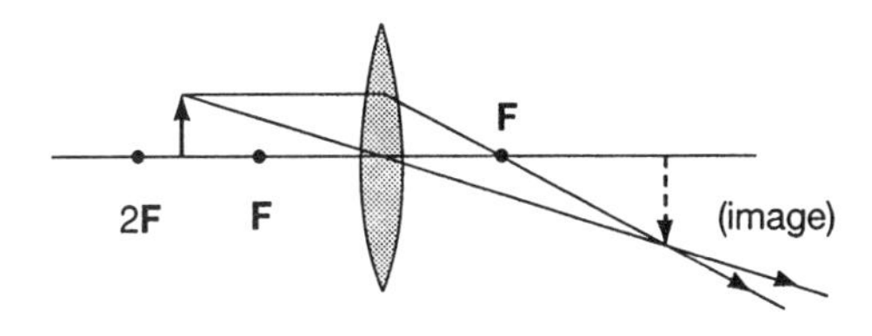

Fig. 22E.46

52. (a) given : $d_i = 400$ cm ; $d_o = 6.0$ cm

$1/f = 1/d_o + 1/d_i$

$1/f = 1/400$ cm + $1/6.0$; $f = 5.9$ cm

(b) $M = -d_i / d_o = -400$ cm / 0.060 cm = -67

$h_i = (67)(1.0$ cm$) = 67$ cm

61. (a) $f = 0.045$ m ; $d_o = 1.5$ m

$1/f = 1/d_o + 1/d_i$

$1/0.045$ m = $1/1.5$ m + $1/d_i$; $d_i = 0.046$ m = 4.6 cm

(b) $h_i / h_o = d_i / d_o$

$h_i = (26$ cm$)(0.046$ m$) / (1.5$ m$) = 0.80$ cm

70. $P = +1.5$ D ; $P = 1/f$; $1.5 = 1/f$; $f = 0.67$ m

$1/f = (n - 1) [(1/R_1) - (1/R_2)]$

$1/0.67$ m = $(1.6 - 1.0) [(1/0.20$ m$) - (1/R_2)]$

$R_2 = 21$ cm ; $2.5 = (1/0.20$ m$) - (1/R_2)$; $R_2 = 40$ cm

Chapter 22

Sample Quiz

Multiple Choice. Choose the correct answer.

__ 1. A person stands 6.0 m in front of a plane mirror. What is the distance between the object and the image ?
A. 0 m B. 3.0 m C. 6.0 m D. 12 m

__ 2. A concave mirror has a focal length of 4.0 cm. What is the radius of curvature ?
A. 2.0 cm B. 4.0 cm C. 8.0 cm D. 12 cm

__ 3. An object is placed in front of a mirror and the image is real and enlarged. What type of mirror is used ?
A. concave B. convex C. plane

__ 4. An object is placed in front of a mirror and the image is virtual and enlarged. What type of mirror is used ?
A. concave B. convex C. plane

__ 5. An object is placed in front of a lens and a virtual image is formed which is reduced in size. What type of lens is used ?
A. concave B. convex C. plane

__ 6. A lens has a focal length of 20 cm. What is the power of the lens in diopters ?
A. 0.050 D B. 0.5 D C. 5.0 D D. 50 D

__ 7. What are the characteristics of an image formed when a paper is read using a magnifying glass ?
A. real and reduced
B. virtual and reduced
C. real and enlarged
D. virtual and enlarged

Problems

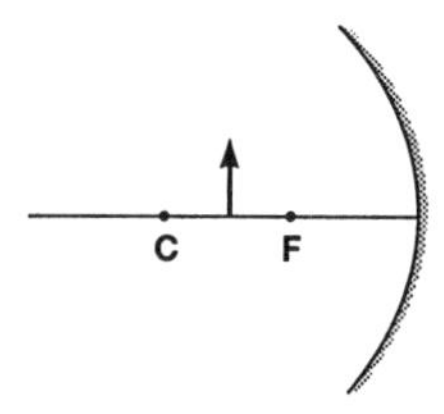

Fig. 22.1

8. A 3.0 cm tall object is placed 6.0 cm from a concave mirror whose radius of is curvature is 8.0 cm.
   A. On the diagram Fig. 22.1, locate the position of the image using two principle rays of light.
   B. Using the mirror equation, find the location of the image.
   C. What is the height of the image ?

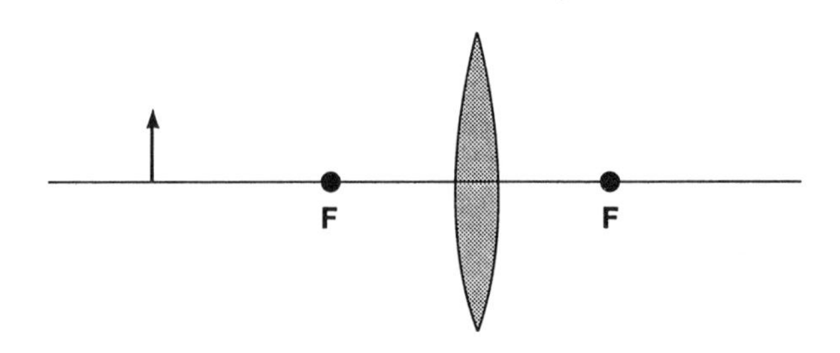

Fig. 22.2

9. A 2.0 cm tall object is located 5.0 cm from a convex lens whose focal length is 2.0 cm.
   A. On the diagram in Fig. 22.2, locate the image using two principle rays of light.
   B. Using the lens equation, find the location of the image.
   C. What is the height of the image?

10. A bioconvex lens is formed by using a piece of plastic (n = 1.7). The radius of curvatures of the two sides are 20 cm and 30 cm. What is the radius of curvature of the combination?

CHAPTER 23 Physical Optics: The Wave Nature of Light

Chapter Objectives

Upon completion of the unit on physical optics, students should be able to :

1. solve problems about and explain double slit interference.
2. explain the interference of thin films, and solve problems on thin film interference.
3. describe the interference effect of Newton's rings.
4. solve problems and explain the characteristics of a single slit diffraction pattern and of patterns formed by diffraction gratings.
5. describe polarization, and explain different ways in which light can be polarized.
6. explain the reasons for optical activity.
7. describe the effect of the scattering of light upon the atmosphere.

Chapter Summary

*__Young's double slit experiment__ demonstrated the wave nature of light and was used to measure wavelength. Interference fringes are produced on a screen as a result of interference because of varying path differences.

*__Thin film interference__ depends on the phases of the light reflected from the film surfaces. (1) A light wave reflected from the boundary of a medium whose index of refraction is greater than that of the incident medium undergoes a 180° phase change. (2) If the reflecting medium has a smaller index of refraction, there is no phase change. For a particular film thickness, light reflected from the film surfaces may interfere constructively or destructively.

*The deviation or bending of light around objects or edges or corners is called **diffraction**. This results in diffraction interference patterns for light passing through slits. For a single slit of width d, the greater the wavelength of light, the wider the diffraction pattern. For a given wavelength, the narrower the slit width, the wider the diffraction pattern. The width of the central maximum is twice the width of the other fringe maxima.

*A large number of parallel slits forms a **diffraction grating**, which can be used to separate the component wavelengths of light like a prism. Diffraction gratings are widely used in spectroscopy. The atomic spacings in crystalline solids act as a diffraction grating for X-rays. The deviation angle - wavelength relationship in this case is known as <u>Bragg's law</u>.

***Polarization** refers to the orientation of transverse oscillations or field vectors of electromagnetic waves. Light with some partial preferential orientation of the field vectors is said to be partially polarized. If the field vectors oscillate in one plane, the light is then plane or linearly polarized.

*Light can be polarized in several ways. These include:

(a) Reflection. Light that is partially reflected and partially transmitted at a medium surface is partially polarized. The maximum polarization occurs when the reflected and refracted beams are 90° apart. The incident angle for this maximum polarization is called the polarizing or Brewster's angle.

(b) Polarization by double refraction (birefringence and dichroism). In some crystalline materials, anisotropy of the speed of light with direction gives rise to different indices of refraction in different directions. Such materials are said to be birefringent or doubly refracting. When a beam is refracted, we find that the two rays are linearly polarized in mutually perpendicular directions. Some birefringent crystals exhibit the property of selectively absorbing one of the polarized components more that the other. This property is called dichroism. Tiny dichroic crystals formed the basis of the original Polaroid film. Now Polaroid film uses the orientation of long polymer chains to effect light polarization.

*Some transparent materials have the ability to rotate the polarization direction of the linearly polarized light. This is called optical activity and is due to the molecular structure of the material. Some liquid crystals are optically active, and this property forms the basis of the common liquid crystal displays (LCD's).

*Scattering is the process of particles (like molecules of the air) absorbing and reradiating light. The scattering of sunlight by air molecules causes the sky to look blue. **Rayleigh scattering** is the scattering found to be proportional to $1/\lambda^4$. and is preferential, with light in the blue end of the visible spectrum being scattered more than that in the red end. The scattering of sunlight by atmospheric gases and small foreign particles give rise to red sunsets.

Important Terms and Relationships

23.1 Young's Double Slit Experiement

physical (wave) optics
Young's double-slit experiment
path difference for constructive interference : $\Delta = n\lambda$, for n = 1,2,3, ...
path difference for destructive interference : $\Delta = m\lambda/2$, for m = 1,2,3, ...
bright fringe condition (double-slit) : $d \sin\theta = n\lambda$, for n = 0,1,2,3, ...

wavelength measurement (double slit) : $\lambda = y_n d / nL$, for n = 1,2,3, ...

width of bright fringe : $y_{n+1} - y_n = L\lambda / d$

23.2 Thin-Film Interference

thin-film interference
optical flats
Newton's rings
non-reflecting film thickness : $t = \lambda / 4n$

23.3 Diffraction

diffraction
diffraction grating
dark fringe conditions (single-slit diffraction) ; $w \sin\theta = m\lambda$, ($m = 1,2,3, \ldots$)
lateral displacement of dark fringe : $y_m = mL\lambda / w$ (for 1,2,3, ...)
width of bright fringe $\Delta = y_{m+1} - y_m = L\lambda / w$
interference maxima for a grating : $d \sin\theta = n\lambda$ for $n = 0,1,2,3, \ldots$
limit of the order number : $n \le d / \lambda$
Bragg's law : $2 d \sin\theta = n\lambda$, for $n = 1,2,3, \ldots$

24.4 Polarization

polarization
polarizing (Brewster's) angle : $\tan\theta = n$
bifringence
dichroism
transmission axis (polarization direction)
optical activity
LCD (liquid crystal display)

23.5 Atmospheric Scattering of Light

scattering
Raleigh scattering

Additional Solved Problems

23.1 Young's Double Slit Experiement

Example 1 A beam of red light having a wavelength of 630 nm passes through two slits which are separated by a distance of 1.0×10^{-2} mm. An interference pattern is located on a screen 2.0 m from the slits.

A. What is the angle between a first order image and the center of the central image ?

B. How far is the first order image located from the center of the central image.

given : $\lambda = 6.30 \times 10^{-7}$ m ; $d = 1.0 \times 10^{-5}$ m ; $x = 2.0$ m

(a) $n \lambda = d \sin \theta$

$(1)(6.30 \times 10^{-7} \text{ m}) = (1.0 \times 10^{-5} \text{ m})(\sin \theta)$; $\theta = 3.6°$

(b) $\tan \theta = y / x$; $\tan 3.6° = y / (2.0 \text{ m})$; $y = 0.13$ m

Example 2 Light passes through fringes which are separated by a distance of 5.0×10^{-6} m. A third order bright is located a distance of 1.5 m from the slits and a distance of 10 cm from center of the central image.

A. What is the wavelength of the light ?

B. Is the light visible ?

given : $d = 5.0 \times 10^{-6}$ m ; $n = 3$; $y_n = 0.10$ m ; $L = 1.5$ m

(a) $\lambda = y_n d / nL$; $\lambda = (10 \times 10^{-2} \text{ m})(5.0 \times 10^{-6} \text{ m}) / (3)(1.5 \text{ m}) = 1.1 \times 10^{-7}$ m

(b) no, the light is not in the range of visible light --- 4.0×10^{-7} m to 7.0×10^{-7} m

23.2 Thin-Film Interference

Example 1 A soap bubble of thickness, t, appears red in a spot. Estimate the minimum thickness of the bubble in the red area. Take the wavelength of red light as 650 nm

given : $\lambda = 650$ nm ; $n = 1.33$

The bubble is mostly water therefore the index of refraction is 1.33. The ray which strikes the top surface must be shifted by a half - wavelength or 325 nm.

To get constructive interference, the second ray must move at least an extra half wavelength to be in phase.

$2t = \theta' / 2$ where λ' is the wavelength in the water.

$\lambda' = \lambda / n$; so $t = \lambda / 4n = (650 \times 10^{-9}) / 4(1.33) = 1.2 \times 10^{-7}$ m

Example 2 A glass is coated with a thin layer of a material which has an index of refraction of 1.8. This makes the lens nonreflecting for light with a wavelength of 500 nm (in air) that is normally incident of the lens. What is the thickness of the thinnest film that will make the lens nonreflecting ?

given : $n = 1.8$; $\lambda = 500$ nm

The minimum thickness for interference to occur is $\lambda' / 4$, where λ' is the wavelength of the light in the film. Light undergoes a 180° phase shift on reflection from each surface and $2t = \lambda' / 2$; $t = \lambda_2 / 4 = \lambda_1 / 4n$

$t = (500 \times 10^{-9}\ \text{m}) / 4(1.8) = 6.9 \times 10^{-8}$ m

23.3 Diffraction

Example 1 A beam of light has a wavelength 500 nm passes through a diffraction grating which has 530 lines/mm. What is the distance of a second order image from the central image on a screen which is 1.0 m from the grating ?

given : $d = 1/530\ \text{mm} = 1.89 \times 10^{-3}\ \text{mm} = 1.89 \times 10^{-6}$ m ; $\lambda = 5.00 \times 10^{-7}$ m

$n\lambda = d \sin\theta$; $(2)(5.00 \times 10^{-7}\ \text{m}) = (1.89 \times 10^{-6}\ \text{m}) \sin\theta$; $\theta = 31.9°$

$\tan\theta = x / y$; $\tan 31.9° = x / 1.0$ m ; 0.62 m

Example 2 Light passes through a diffraction grating which has 400 lines/mm. It is observed the second order image is located a distance of 12.0 cm from the central bright spot on a screen which is 120 cm from the grating. What is the wavelength of the light ?

first find $d = 1/400 = 2.5 \times 10^{-3}$ mm $= 2.5 \times 10^{-6}$ m

also given : $n = 2$; $x = 0.120$ m ; $y = 1.20$ m

next find θ : $\tan\theta = x / y = 0.120 \text{ m} / 1.20 \text{ m}$; $\theta = 5.71°$
$d \sin\theta = n \lambda$
$(2.50 \times 10^{-6} \text{ m}) (\sin 5.71°) = (2) \lambda$; $\lambda = 1.2 \times 10^{-7}$ m

Example 3 Light passes through a slit, width 9.0×10^{-2} mm has a second order dark fringe 1.0 cm from the middle of the central maximum on a screen 1.0 m from the slit. What is the wavelength of the light ?

given : $y_m = 1.0 \times 10^{-2}$ m ; $L = 1.0$ m ; $w = 9.0 \times 10^{-5}$ m ; $m = 2$; find λ

$y_m = m \lambda L / w$

$(1.0 \times 10^{-2} \text{ m}) = (3) (\lambda)(1.0 \text{ m}) / (9.0 \times 10^{-5} \text{ m})$; $\lambda = 4.5 \times 10^{-7}$ m

Example 4 Light, wavelength 580 nm, passes through a slit whose width is 1.00×10^{-2} cm. The diffraction pattern is observed on a screen 1.50 m from the slit.

A. What is the angular width of the central bright ?
B. What is the linear width of the central bright fringe ?

A. given : $\lambda = 580$ nm $= 5.80 \times 10^{-7}$ m ; $w = 1.00 \times 10^{-4}$ m ; $x = 1.50$ m ; $m = 1$

$m \lambda = w (\sin\theta)$; $1.0 (5.80 \times 10^{-7} \text{ m}) = (1.00 \times 10^{-4} \text{ m}) \sin\theta$

$\theta = 0.33°$; $2\theta = 0.66°$

B. $\tan\theta = x / y$; $\tan(0.33°) = x / (1.0 \text{ m})$; 5.8×10^{-3} m
$2x = 1.2 \times 10^{-2}$ m

24.4 Polarization

Example The angle of incidence is adjusted so there is maximum linear polarization for the reflection of light from a transparent piece of plastic, $n = 1.45$. What is the angle of incidence ?

given : $n = 1.45$; find θ_p

$\tan\theta_p = 1.45$; $\theta_p = 55.4°$

Solutions to paried problems and other selected problems

4. Light must be coherent for interference patterns to be observed.

10. given : $d = 0.20$ m $= 2.0 \times 10^{-4}$ m ; $L = 1.5$ m ; $n = 1.0$ (for first bright ; remember there is a bright in the center) ; $y_n = 0.45$ cm $= 4.5 \times 10^{-3}$ m ; find λ .

$\lambda = y_n d / nL$

$\lambda = (4.5 \times 10^{-3}\text{ m})(2.0 \times 10^{-4}\text{ m}) / (1.0)(1.5\text{ m}) = 6.0 \times 10^{-7}$ m or 600 nm
this light is yellow to orange.

18. given : $L = 1.50$ m ; $\theta_3 - \theta_2 = 0.0230$ rad ; $d = 0.0350$ mm $= 3.50 \times 10^{-5}$ m

(a) $\theta = 2\lambda / d$; $0.0230 = 2(\lambda) / (3.50 \times 10^{-5})$; $\lambda = 4.00 \times 10^{-7}$ m
violet light

(b) $n\lambda = y_n d / L$; $(2.0)(4.00 \times 10^{-7}\text{ m}) = y\,(3.50 \times 10^{-5}\text{ m}) / (1.50\text{ m})$

$y = 3.4 \times 10^{-2}$ m

22. the minumim thickness should be (a) $\lambda' / 4$.

27. given : $n = 1.38$; $\lambda_1 = 700 \times 10^{-9}$ m ; $\lambda_2 = 400 \times 10^{-7}$ m ; find t

$t = \lambda / 4n$

$t_1 = (700 \times 10^{-9}\text{ m}) / 4\,(1.38) = 1.27 \times 10^{-7}$ m

$t_2 = (400 \times 10^{-9}\text{ m}) / 4\,(1.38) = 7.2 \times 10^{-8}$ m

$\Delta t = t_1 - t_2 = 5.5 \times 10^{-8}$ m

34. (c) the smaller the opening the greater the spreading ; d and sin θ are inversely

44. given : $d = 1 / 8.0 \times 10^3 = 1.3 \times 10^{-6}$ m ; $\lambda_1 = 400 \times 10^{-7}$ m ; $\lambda_2 = 700 \times 10^{-9}$ m

$n = 1.0$; find θ_1 and θ_2

$n\lambda = d \sin\theta$

$(1.0)(400 \times 10^{-7}\text{ m}) = (1.3 \times 10^{-6}\text{ m})\sin\theta_1$; $\theta_1 = 18°$

$(1.0)(700 \times 10^{-7}\text{ m}) = (1.3 \times 10^{-6}\text{ m})\sin\theta_2$; $\theta_2 = 33°$

$\Delta\theta = 15°$

50. $\lambda_1 = 400 \times 10^{-9}$ m ; $\lambda_2 = 600 \times 10^{-9}$ m ; $n_1 = 3.0$; $n_2 = 2.0$

$d\sin\theta = n\lambda$; $d\sin\theta_1 = (3.0)(400 \times 10^{-9}) = 12 \times 10^{-9}$

$d\sin\theta_2 = (2.0)(600 \times 10^{-9}) = 12 \times 10^{-9}$

since these are equal ; their location is the same and the images overlap

52. light can be polarized by reflection, refraction, and by scattering (d)

56. $\tan\theta_p = n$; $\tan\theta_1 = 1.4$; $\theta_1 = 54°$

$\tan\theta_2 = 1.7$; $\theta_2 = 60°$

62.

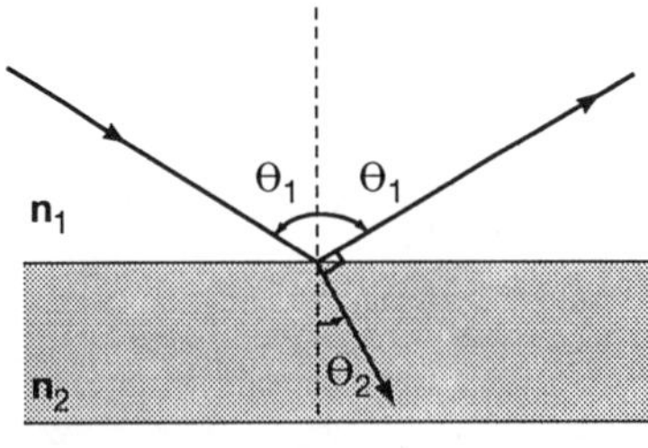

Fig. 23E.62

$\theta_2 = 90° - \theta_1$

$\sin\theta_1 / \sin\theta_2 = n_2 / n_1 = \sin\theta_1 / \cos\theta_1 = \tan\theta_p$

Sample Quiz

Multiple Choice. Choose the correct answer.

__ 1. Light passes through a diffraction grating. As the order image increases, the intensity of the light
A. increases B. decreases C. remains the same

__ 2. Beams of different colored lights passes through two slits. Which of the following colors will be located the greatest distance from the central bright ?
A. yellow B. red C. green D. blue

__ 3. A beam of light passes through a single slit. The width of the slit is decreased. What happens to the width of the central bright fringe ?
A. increases B. decreases C. remains the same

__ 4. White light passes through a diffraction grating. Which color of light is located the greatest distance from the central image ?
A. yellow B. red C. green D. blue

__ 5. A beam of light passes through four different diffraction grating. Which grating will have the least number of bright fringes ?
A. 53 lines/mm B. 530 lines/mm C. 5300 lines/mm D. 8000 lines/mm

__ 6. Which of the following cannot be polartized ?
A. x-rays B. red light C. sound D. infrared light

__ 7. What is Brewster's angle an air to water surface (n = 1.33) ?
A. 49° B. 41° C. 53° D. 47°

Problems

8. A beam of light passes through a slit whose width 1.0×10^{-6} m. The wavelength of the light is 580 nm. What is the linear distance of the central bright finge on a screen 2.0 m from the slit ?

9. White light passes through a diffraction grating which has 530 lines/mm. What is the linear width of a first order image on a screen 2.0 m from the grating ?

10. Sunlight is reflected from the smooth surface of a piece of glass (n = 1.6). What is the Sun's altitude when the polarization of the reflected light is greatest ?

CHAPTER 24 Optical Instruments

Chapter Objectives

Upon completion of the unit on optical instruments, students should be able to :

1. describe the functions of the major parts of the human eye.
2. distinguish between far point and near point, describe the defects of nearsightedness and farsightedness, and give solutions for correcting the defects.
3. list the cause of astigmatism, and qualitatively describe measures to correct the defect.
4. find the magnification for an image at a near point (25 cm) and at infinity.
5. calculate angular magnification.
6. find the magnification for a compound microscope, and explain the purpose of the lenses in the microscope.
7. compare and contrast the two types of optical telescopes, and find the magnification for a refracting telescope.
8. find the minimum angle of resolution for a slit width.
9. find the minimum separation angle for the Rayleigh criterion of two objects, and find the resolving power for a lens.
10. state the theory for color vision.
11. know the three primary colors and how they can be combined to produce other colors.
12. state the three primary pigments, and explain how the subtractive process can be used to obtain other colors.

Chapter Summary

*Through accommodation or the changing of the shape of the eye's crystalline lens through muscle action, sharp images are formed on the retina of the normal eye. The photo-sensitive rod and cone cells of the retina are responsible for twilight vision and color vision, respectively. The extremes of the range over which distant vision is possible are known as the far point and the near point. The near point gradually recedes with age because of a loss of elasticity in the crystalline lens.

*Nearsightedness is the condition of being able to see nearby objects clearly, but not distant objects. This may be corrected by using eyeglasses with an appropriate diverging lens. Farsightedness is the condition of being able to see far objects clearly, but not nearby objects. This may be corrected by using eyeglasses with an appropriate converging lens. Astigmatism, which is usually caused by nonspherical or out-of-round refractive surface(s), is the condition of the eye having different focal planes in different planes. This condition may be corrected by using eyeglasses with a curvature that compensates for the difference of an eye surface.

*A **magnifying glass** or simple microscope is simply a single convex lens which allows us to view clearly an object brought closer than the near point. In doing so, the object is magnified. This is expressed in terms of **angular magnification** , which is the range of the ratio of the angular size of the object viewed through the magnifying glass and the angular size of the object seen without the magnifying glass.

*A **compound microscope** consists of a pair of converging lenses, each of which contributes to the magnification. A converging lens of relatively short focal length, called the **objective**, produces a real, inverted, and enlarged image of a specimen object positioned slightly beyond the focal point of the lens. The other lens, the **eyepiece** or **ocular**, which has a longer focal length, is positioned so that the image formed by the objective falls inside its focal point. In essence, the objective acts as a projector, and the eyepiece is used as a simple microscope to view the projected image.

*Two types of telescopes are : refracting and reflecting. A reflecting telescope uses a converging lens to collect and converge light from a distant object. In a refractive astronomical telescope, which uses converging lens as an eyepiece, the final image is inverted. In a refractive **terrestrial telescope**, the final image is upright. This may be accomplished by using a diverging lens as an eyepiece (Galilean telescope) or a third converging "erecting" lens between the objective and the eyepiece lenses.

*Diffraction places a limitation of our ability to distinguish closely separated objects when using microscopes and telescopes. In general, two sources can be resolved if the central maximum of the diffraction pattern of one fails at or beyond the first minimum (dark fringe) of the other. This generally accepted limiting condition of resolution of two different diffracted images is known as the **Rayleigh criterion**: two images are said to be resolved when the central maximum of one image falls on the first minimum of the diffraction pattern of the other image.

*The **resolving power** of the objective lens of a microscope is the minimum separation distance between two point objects or specimen details that can just be resolved.

*The **additive primary colors** are red, blue, and green. The mixing of light of the additive primaries is called the **additive method of color production**. The proper mixing of the primary colors appear white to the eye. Pairs of color combinations that appear white to the eye are called **complementary colors**. The complement of blue is yellow, of red is cyan, and of green is magenta.

*The **subtractive primary pigments** (subtractive primaries) are cyan, magenta, and yellow. A mixture of absorbing pigments results in the subtraction of colors, and one sees the color of light that is not absorbed or subtracted. This is called **subtractive method of color production**. When three primary pigments are mixed in the proper proportion, the mixture appears black (all wavelengths are absorbed). Painters often refer to the subtractive primaries as being red, yellow, and blue (magenta, yellow, and cyan). By mixing paints of these colors in the proper proportions, a broad spectrum of colors can be produced.

Important Terms and Relationships

24.1 The Human Eye

crystalline lens
retina
rods
cones
accommodation
far point
near point
nearsightedness
farsightedness
astigmatism

24.2 Microscopes

magnifying glass (simple microscope)
angular magnification : $m = \theta / \theta_o$
compound microscope : $M_t = M_o M_e = 25\,L / f_o f_e$
objective
eyepiece (ocular)
magnification of a magnifying glass
with image of near point (25 cm) : $m = 1 + (25\text{ cm} / f)$
magnification of a magnifying glass with image at infinity : $m = 25\text{ cm} / f$

24.3 Telecsopes

refracting telescope
magnification : $m = f_o / f_e$

astronomical telescope
terrestrial telescope
reflecting telescope

24.4 Diffraction and Resolution

resolution
for a slit : $\theta_{min} = \lambda / d$

for a circular aperature : $\theta_{min} = 1.22\,\lambda / D$

Rayleigh criterion
resolving power : $s = f\,\theta_{min} = 1.22\,\lambda f / D$

25.5 Color and Optical Illusions

additive primary colors
additive method of color production
complementay colors
subractive methods of color production
subtractive primaries

Additional Solved Problems

24.1 The Human Eye

Example A farsighted person with a near point of 100 cm gets contact lenses and can read a newspaper held at a distance of 25 cm. What is the power of the lenses ?

given : d_o = 25 cm ; d_i = -100 cm ; find f

$1/f = 1/d_o + 1/d_1$

$1/f = 1 / 25$ cm + 1/(- 100 cm) ; f = 33.3 cm = 0.333 m

P = 1 / f ; P = 1/0.33 = +3.0 D

24.2 Microscopes

Example A microscope has an objective lens with a focal length of 4.00 mm and eyepiece with a focal length of 10.0 mm. The lenses are 20.0 cm apart. Determine the magnification.

given : $f_o = 0.400$ cm ; $f_e = 1.00$ cm ; $L = 20.00$ cm

$M_t = (25 \text{ cm}) L / f_o f_e$

$M_t = (25 \text{ cm})(20.0 \text{ cm}) / (0.40 \text{ cm})(1.00 \text{ cm}) = 1.25 \times 10^3$ X

24.3 Telecsopes

Example In a terrestrial telescope, an objective lens has a focal length of 50 cm and the eyepiece has a focal length of 20 cm. The focal length of the erecting lens is 10 cm. What should be the overall length of the telescope ?

given : $f_o = 50$ cm ; $f_e = 10$ cm ; $f = 10$ cm

$L_t = 4f + L = 4f + (f_e + f_o)$

$L_t = 4 (10 \text{ cm}) + [(20 \text{ cm}) + (50 \text{ cm})] = 110$ cm

24.4 Diffraction and Resolution

Example 1 According to the Rayleigh criterion, what is the minimum angle of resolution for two point sources of yellow light (wavelength 540 nm) produced by a slit width of 0.25 mm ?

$\lambda = 5.40 \times 10^{-7}$ m ; $d = 0.25 \times 10^{-3}$ m
$\theta_{min} = \lambda / d = (5.40 \times 10^{-7}\ m) / (0.25 \times 10^{-3}\ m) = 2.2 \times 10^{-3}$ rad

Example 2 What is the resolution limit due to diffraction for a telescope 100 cm long with a wavelength of 500 nm ?

given : $\lambda = 5.00 \times 10^{-7}$ m ; D = 1.00 m
$\theta_{min} = 1.22\ \lambda / D$; $\theta_{min} = (1.22)(5.00 \times 10^{-7}\ m) / (1.00\ m)$;
$\theta_{min} = 6.1 \times 10^{-7}$ rad

Solutions to paired problems and other selected problems

2. As one ages, the elasticity of the crystalline lens decreases and a person becomes farsighted.

7. given : $d_o = 0.25$ m ; $d_i = 0.015$ m
first find the focal length :

$1/f = 1/d_o + 1/d_i$; $1/f = (1/0.25\ m) + (1/0.015\ m)$; $f = 1.4 \times 10^{-2}$ m
now find the power :

$P = 1 / f = 1 / (1.4 \times 10^{-2}\ m) = +71$ D

12. given : P = +1.25 D ; $d_o = 0.25$ m
$+1.25 = 1/f$; $f = 0.80$ m
$1/f = 1/d_o + 1/d_i$
$(1/0.80\ m) = (1/0.25\ m) + 1/d_i$; $d_i = -0.36$ m

22. given : f = 16 cm ; find m
$m = 1 + (25\ cm / f)$
$m = 1 + (25\ cm / 16\ cm) = 2.6$ X

26. given : f = 15 cm ; find m
(a) m = 1 + (25 cm / f)
m = 1 + (25 cm / 15 cm) = 2.7 X
(b) m = (25 cm / f) = (25 cm / 15 cm) = 1.7 X

32. given : L = 16 cm ; f_e = 3.0 cm ; f_o = 0.45 cm ; find M_t
M_t = (25 cm)L / ($f_o f_e$)
M_t = (25 cm)(16 cm) / [(3.0 cm)(0.45 cm) = 3.0 x 10^2 X

40. (d)

46. given : f_o = 87.5 cm ; f_e = 0.750 cm ; find M
(a) m = f_o / f_e = (87.5 cm) / (0.750 cm) = 117 X
(b) L = f_e + f_o = 87.5 cm + 0.750 cm) = 88.3 cm

52. (d) reduce the wavelength of light so as to increase the resolving power. (refraction)

56. given : λ = 550 x 10^{-9} m ; D = 1.02 m ; find θ_{min}
θ_{min} = (1.22) λ / D ; θ_{min} = (1.22)(550 x 10^{-9} m) / (1.02 m) = 6.58 x 10^{-7} rad

58. λ = 570 x 10^{-7} cm ; f = 3.0 cm ; D = 2.5 cm
(a) θ_{min} = (1.22) λ / D
θ_{min} = (1.22)(5.7 x 10^{-5}) / 2.5 = 2.8 x 10^{-5} rad
(b) s = f θ_{min} = (3.0)(2.8 x 10^{-5} cm) = 8.4 x 10^{-5} cm = 8.4 x 10^{-4} mm

64. red : the white regions would reflect red and the red region would reflect red ; the blue region contains no red, therefore this light is abosorbed and appears black.
blue : the blue and white are reflected blue and the red regions appears black.
green : only the white region reflects the green light ; the rest of the light is absorbed.

Sample Quiz

Multiple Choice. Choose the correct answer.

__ 1. Which of the following is not a primary color of light ?
A. blue B. green C. red D. yellow

__ 2. What is the complement of yellow ?
A. blue B. green C. red D. magenta

__ 3. When the pigments of cyan and yellow are combined, the color produced is
A. blue B. green C. red D. orange

__ 4. Which color of light would have the best resolving power ?
A. blue B. green C. red D. yellow

__ 5. The greatest distance at which the normal eye can see objects clearly is
A. near point B. far point C. diverging D. converging

__ 6. Which of the following visions is best ?
A. 20/20 B. 15/20 C. 10/20 D. 20/15

__ 7. What is the complement of cyan ?
A. green B. blue C. red D. yellow

Chapter 24

Problems

8. A nearsighted person wears glasses whose lenses have a power of -0.15 D. What is the person's far point ?

9. A compound miscroscope has an 18 cm barrel and an ocular with a focal length of 8.0 mm. What is the focal length of the objective to give a total magnification of 360 X ?

10. A refracting telescope has an objective with a focal length of 40 cm and eyepiece with a focal length of 15 mm. The telescope is used to view an object that is 10 cm high and located 100 m away. What is the apparent height of the object as viewed through the telescope ?

CHAPTER 25 Relativity

Chapter Objectives

Upon completion of the unit on relativity, the students should be able to :

1. describe effects of classical relativity.
2. explain Newton's first law of motion in terms of inertial reference frames.
3. explain the significance of the work or major scientists, including Einstein, Michelson, Morley, and Lorentz.
4. explain the effects of the speed on fundamental quantities of mass, length, and time.
5. calculate the speeds of objects from relativistic effects.
6. sketch graphs of the mass, time, length, momentum, and kinetic energy of an object as it approaches the speed of light.
7. write and apply expressions for relativistic length, time, and mass.
8. write and apply expressions for relativistic momentum, kinetic energy, and total energy.
9. explain the principle of the general theory of relativity.

Chapter Summary

*An **inertial frame of reference** is one in which Newton's first law of motion holds. In an inertial system, an isolated body would be stationary or moving with a constant velocity. The **principle of classical** or **Newtonian relativity** states: the laws of mechanics are the same in all inertial reference frames.

*Classically, through vector addition, the speed of light should be referenced to a particular frame. This gives rise to a unique reference frame associated with the proposed medium of transport of the electromagnetic waves called **ether**. As a result, it would seem that Maxwell's equations did not satisfy the Newtonian principle as did other physical laws.

*The **Michelson-Morley experiment** was an attempt to detect the ether by using the interference of light in an extremely sensitive <u>interferometer</u>. If the ether existed, interference fringe shifts should be observed, but the experiment had <u>null</u> results. Attempts were made to explain the result in terms of ether "drag" and material (Lorentz-Fitzgerald) contraction.

*The null result of the Michelson-Morley experiment, or the failure to detect the ether, was resolved by the **special theory of relativity** introduced by Albert Einstein. This theory had two basic postulates :

Postulate I (Principle of Relativity) : The laws of physics are the same in all inertial reference frames.

Postulate II (Constancy of the Speed of Light): The speed of light c in vacuum is constant in any inertial reference frame.

*The special theory of relativity predicts an effect known as time dilation. **Time dilation** refers to an observer noting that a clock in a moving system relative to his own runs slower. Since the effect is relative, to distinguish between the two time intervals, we refer to **proper time** t_o, which is the time interval measured between two events in the reference frame where the events occur at the same location or point in space, and $t = \gamma t_o$, where t is the relative time and $\gamma = [\,1 - (v/c)^2]^{1/2}$.

*The special theory of relativity predicts an effect known as length contraction. **Length contraction** refers to an observer noting that a length L in a moving system, relative to his own is shorter than an equivalent length in his system. Since the effect is relative, to distinguish between the two length intervals, we refer to the **proper length** L_o, which is the length measured by an observer in the length rest frame, and $L = L_o/\gamma$

*Time dilation gives rise to the **twin or clock paradox**. A twin on a space journey relative to Earth would age more slowly than the other Earth-bound twin.

*Other ramifications of the theory of relativity are relativistic mass and relativistic energy. The **relativistic mass** m of a particle increases with speed, $m = \gamma m_o$, where m_o is the rest (or proper) mass. As a result, the **relativistic momentum** is $p = mv = \gamma m_o v$. The **relativistic kinetic energy** is given by $K = mc^2 - m_o c^2$, or $K = E - E_o$, where $E_o = m_o c^2$ is the **rest energy**. This is the famous **mass-energy formula**, which points out that mass is a form of energy.

*The **general theory of relativity** considers non-inertial or accelerating systems. It is basically a gravitational theory. Its **principle of equivalence** states: an inertial reference frame in a uniform gravitational field is equivalent to a reference frame that has a constant acceleration with respect to the inertial frame. What this basically means is that : no experiment performed in a closed, accelerating system can distinguish between the effects of a gravitational field and the effect of the acceleration.

*The general theory of relativity predicts such effects as the gravitational bending of light, the view of gravitational field as a "warping" of times and space, and the gravitational red shift of light.

Important Terms and Relationships

25.1 Classical Relativity

relativity
inertial reference frame
principle of classical (Newtonian) relativity

25.2 The Michelson-Morley Experiment

ether
Michelson-Morley experiment
inferometer

25.3 The Special Theory of Relativity

principle of relativity
special theory of relativity
constancy of the speed of light
time dilation : $t = \gamma t_o = t_o / [1 - (v/c)^2]^{1/2}$
proper time
proper length
length contraction : $L = L_o / \gamma = L_o / [1 - (v/c)^2]^{1/2}$

25.4 Relativistic Mass and Energy

rest mass
relativistic mass : $m = \gamma m_o = m_o / [1 - (v/c)^2]^{1/2}$
relativistic momentum : $p = mv = \gamma m_o v$
relativistic kinetic energy : $K = E - E_o = (m - m_o) c^2$
total energy : $E = K + E_o$; $mc^2 = K + m_o c^2$; $E = \gamma E_o^2$

rest energy : $E = m_o c^2$

mass-energy equivalence

25.5 The General Theory of Relativity

general theory of relativity
principle of relativity
black hole

Schwarzschild radius : $R = 2GM / c^2$
event horizon

25.6 Relativistic Velocity Addition $u = v + u' / [1 + vu'/c^2]$

Additional Solved Problems

25.1 Classical Relativity

Example A train is traveling at 25 m/s. A car is traveling at 30 m/s on a straight track beside the train.

A. If the car is traveling in the same direction as the train, how long will take the car to catch up with the train if the two are separated by a distance of 500 m ?

B. If the car is 1000 m from the train and the car is traveling in the opposite direction of the train, how long will it take the car to meet the train ?

given : $v_1 = 30$ m/s ; $v_2 = 25$ m/s

(a) $v_r = v_1 - v_2 = 30$ m/s - 25 m/s = 5 m/s relative to the train

$d = vt$; 500 m = (5 m/s) t ; t = 100 s

(b) $v_r = v_1 + v_2 = 30$ m/s + 25 m/s = 55 m/s

1000 m = (55 m/s) t ; t = 18 s

25.2 The Michelson-Morley Experiment

Example An object is moving toward you with a speed of 0.80 c. By what percent will its length appear to be as compared to the object at rest ?

given : $v = 0.80\ c$

$\gamma = 1 / [1 - v^2 / c^2]^{1/2}$

$\gamma = 1 / [1 - (0.80)^2]^{1/2}$

$\gamma = 1.67$

$L_o = \gamma L = 1.67\ L$; $L = 60\%(L_o)$

25.3 The Special Theory of Relativity

Example 1 A spaceship is traveling with a speed of 0.60 c. It appears to be 10.0 m long. How long will it appear to be if the ship were at rest ?

given : $v = 0.60\ c$; $L = 10.0\ m$

$\gamma = 1 / [1 - v^2 / c^2]^{1/2}$

$\gamma = 1 / [1 - (0.60)^2]^{1/2}$

$\gamma = 1.25$

$L_o = L\gamma$; $L_o = 1.25\ (10.0\ m) = 12.5\ m$

Example 2 A certain particle has a lifetime of 5.0×10^{-10} s at rest. What is the observed lifetime of the particle if it travels with a speed of 0.80 c ?

given : $t_o = 5.0 \times 10^{-10}\ s$; $v = 0.80\ c$

$\gamma = 1 / [1 - v^2 / c^2]^{1/2}$

$\gamma = 1 / [1 - (0.80)^2]^{1/2}$

$\gamma = 1.67$

$t = \gamma t_o = (1.67)(5.0 \times 10^{-10}\ s) = 8.4 \times 10^{-10}\ s$

Example 3 Alpha Centuri, the binary star closest to this solar system, is about 4.30 light years away. A spaceship with a constant velocity of 0.80 c relative to Earth travels toward the system.

A. How much time would elapse on a clock on the spaceship ?
B. What would the occupants on the spaceship measure the distance to be to Alpha Centuri ?

given : $v = 0.80\ c$; $L_o = 4.3$ ly

first find earth time ; $d = vt$; $(4.3\ ly) = (0.80\ c)\ t$; $t = 5.4$ y

(a) $\gamma = 1 / [1 - v^2 / c^2]^{1/2}$

$\gamma = 1 / [1 - (0.80)^2]^{1/2}$

$\gamma = 1.67$

$t_o = t / \gamma = 5.4 / 1.67 = 3.2$ y

(b) $L_o = \gamma L$; $4.3\ ly = (1.67)\ L$; $L = 2.6$ ly

25.4 Relativistis Mass and Energy

Example 1 A proton is traveling with a speed of 0.80 c. What is its relativistic mass ?

given : $v = 0.80\ c$; $m_o = 1.67 \times 10^{-27}$ kg

$\gamma = 1 / [1 - v^2 / c^2]^{1/2}$

$\gamma = 1 / [1 - (0.80)^2]^{1/2}$

$\gamma = 1.67$

$m = \gamma m_o = (1.67)(1.67 \times 10^{-27}\ kg) = 2.8 \times 10^{-27}$ kg

Example 2 The kinetic energy of an electron is 80% of its total energy.
A. What is the speed of the electron ?
B. What is the momentum of the electron ?

(a) $E = K + E_o$

$E = (0.80)E + E_o$; $0.20\ E = E_o$

$E = \gamma E_o$; $\gamma = 5.0$

$\gamma = 1 / [1 - v^2 / c^2]^{1/2}$

$5.0 = 1 / [1 - v^2 / c^2]^{1/2}$

$1 - v^2 / c^2 = 0.04$

$1 - 0.04 = v^2 / c^2$; $v = 0.98\ c$

(b) $p = \gamma m_o v$

$p = (5.0)(1.67 \times 10^{-27}\ kg)(0.98\ c) = 2.5 \times 10^{-18}$ kg-m/s

Example 3 A proton moves with a speed of 0.80 c. What is its
A. total energy ?
B. kinetic energy ?
C. relativistic momentum ?

given : $v = 0.90\ c$; $m_o = 1.67 \times 10^{-27}$ kg

first find : $\gamma = 1 / [\ 1 - v^2/c^2\ ]^{1/2}$; $\gamma = 1 / [\ 1 - (0.80\ c)^2 / c^2]^{1/2}$; $\gamma = 1.67$

(a) $E = mc^2$; $E = \gamma m_o c^2 = (1.67)(1.67 \times 10^{-27}\ \text{kg})(3.0 \times 10^8)^2 = 2.5 \times 10^{-10}$ J

(b) $K = m_o c^2\ (\gamma - 1) = (1.67 \times 10^{-27}\ \text{kg})(3.0 \times 10^8\ \text{m/s})^2 (1.67 - 1) = 1.0 \times 10^{-10}$ J

(c) $pc = (E^2 - E_o{}^2)^{1/2}$

$p\ (3.0 \times 10^8\ \text{m/s}) = \{\ (2.5 \times 10^{-10}\ \text{J})^2 - [(1.67 \times 10^{-27}\ \text{kg})(3.0 \times 10^8)]^2\}^{1/2}$

$p = 8.3 \times 10^{-19}$ kg-m/s

25.5 The General Theory of Relativity

Example A hypothetical black hole has a radius of 10 m. What would be the mass of the black hole ?

given : R = 10 m

$c^2 = 2GM/R$

$(3.0 \times 10^8\ \text{m/s})^2 = 2\ (6.67 \times 10^{-11}\ \text{N-m}^2/\text{kg}^2)(M) / (10\ \text{m})$

$M = 6.7 \times 10^{27}$ kg

25.6 Relativistic Velocity Addition

Example A spaceship moves away from the Earth with a speed of 0.80 c. The spaceship then fires a missle with a speed of 0.50 c relative to the ship.
A. What is the relative speed of the missle relative to the Earth, if the missle were fired away from the Earth ?
B. What is the relative speed of the missle relative to the Earth, if the missle were fired toward the Earth ?

given : $u' = 0.50\ c$ with respect to spaceship ; $v = 0.80\ c$ with respect to Earth

(a) $u = v + u' / [\ 1 + (vu' / c^2)]$

$u = (\ 0.80\ c + 0.50\ c) / [\ 1 + (0.80\ c)(0.50\ c) / c^2] = 0.93\ c$

(b) $u = v + u' / [1 + (vu' / c^2)]$

$u = (0.80\ c + -0.50\ c) / [1 + (0.80\ c)(-0.50\ c) / c^2] = 0.50\ c$ with respect to Earth

Solutions of paired problems and other seleted problems

2. (a)

10. given : $v_{plane} = 200$ km/h ; (a) $v_{wind} = -35$ km/h ; (b) $v_{wind} = 25$ km/h

(a) $v_r = v_{plane} - v_{wind} = 200$ km/h - 35 km/h = 165 km/h

(b) $v_r = v_{plane} + v_{wind} = 200$ km/h + 25 km/h = 225 km/h

16. (a) applies to inertial systems

24. given : $L_o = 4.30$ y ; $v = 0.60$ c ; find : t_o

Earth time : $d = vt = (0.6\ c)(4.3\ y) = 7.2$ y ; $\gamma = 1 / [1 - (v/c)^2]^{1/2}$;

$\gamma = 1 / [1 - (0.6\ c / c)]^{1/2} = 1.25$;

$t = \gamma t_o$; $7.2 = t_o (1.25)$; $t_o = 5.7$ y

28. given : $L_o = 35.0$ m ; $h_o = 8.25$ m ; $v = 2.44 \times 10^8$ m/s

$\gamma = 1 / [1 - (v/c)^2]^{1/2}$; $\gamma = [1 - (2.44 \times 10^8 / 3.0 \times 10^8)]^{1/2} = 1.7$

$L = L_o / \gamma =$ (35.0 m) / (1.7) = 20 m ; height (diameter does not change) - 8.25 m

46. given : K = 60% E

$E = K + E_o$

$E = 0.60\ E + E_o$; $E_o = 0.40\ E$

$1/\gamma = 0.40 = [1 - (v/c)^2]^{1/2}$

$v = 0.92$ c

$p = mv = (2.5)(9.11 \times 10^{-31}\ kg)(0.92)(3.0 \times 10^8\ m/s) = 6.3 \times 10^{-22}$ kg-m/s

52. given : $P = 3.827 \times 10^{26}$ W ; $t = 3600$ s

(a) $P = E / t$

3.827×10^{26} W = E / (3600 s) ; $E = 1.378 \times 10^{30}$ J

$E = mc^2$; 1.378×10^{30} J = m $(3.0 \times 10^8 \text{ m/s})^2$; $m = 1.531 \times 10^{13}$ kg

(b) 1.531×10^{13} kg/h = $(1.989 \times 10^{28}$ kg) / x

$x = 1.299 \times 10^{15}$ h = 1.483×10^{11} y

60. (d)

64. given : R = 5.00 m ; $G = 6.67 \times 10^{-11}$ N-m^2 / kg^2 ;

(a) $R = 2GM / c^2$; (5.00 m) = 2 $(6.67 \times 10^{-11}$ N-m^2 / kg^2)(m) / $(3.0 \times 10^8 \text{ m/s})^2$

$R = 3.37 \times 10^{27}$ kg

(b) $\rho = m / V$; $V = (4/3)\pi r^3 = (4/3)(\pi)(5.0 \text{ m})^3 = 5.2 \times 10^2 \text{ m}^3$

$\rho = (3.37 \times 10^{27} \text{ kg}) / (5.2 \times 10^2 \text{ m}^3) = 6.44 \times 10^{24} \text{ kg/m}^3$

68. given : v = 0.40 c ; u' = -0.15 c (opposite direction of ship)

$u = v + u' / (1 + vu' / c^2)$

$u = [(0.40 c) - (0.15 c)] / [1 + (0.40 c)(-0.15 c) / c^2] = 0.27 c$

Sample Quiz

Multiple Choice. Choose the correct answer.

__ 1. What is the rest energy of an electron ?

A. 2.7×10^{-22} J B. 8.2×10^{-14} J

C. 2.5×10^{-5} J D. none of the choices

__ 2. A clock is placed aboard the space shuttle. After a trip traveling at very fast speed, how does a clock aboard the ship compare with a clock on Earth ?

A. less time B. more time C. the same amount of time

Questions 3 - 5 refer to the following.

A spaceship, length 10 m and height 10 m moves toward you with a speed of 0.80 c.

__ 3. What is the apparent length of the spaceship ?

A. 4.0 m B. 6.0 m C. 10 m D. 17 m

__ 4. What is the apparent height of the spaceship ?

A. 4.0 m B. 6.0 m C. 10 m D. 17 m

__ 5. When 10 s is observed on a clock on the ship, how long does a stationary observer notice passes on the clock ?

A. 4.0 s B. 6.0 s C. 10 s D. 17 s

__ 6. As the speed of a particle approaches the speed of light, the mass

A. decreases B. increases C. remains the same

__ 7. A fast car is traveling with a speed of 0.80 c. How fast would light travel from the headlights of the car relative to a stationary observer ?

A. 1.8 c B. c C. 0.80 c D. neither choice

Problems

8. Your friend travels in a car which has a speed of 0.80 c. When at rest the length of the car is 3.0 m.
   A. What is the apparent length of the car relative to a stationary observer ?
   B. When your friend observes 30 s on his watch, how long does it appear to a stationary observer ?

9. A space traveler moves away from the Earth with a speed of 0.80 c when a missile is fired with a speed of 0.40 c relative to the ship away from the Earth. How fast does the missile appear to travel relative to the Earth ?

10. An electron is moving with a speed of 0.80 c.
    A. What is the momentum of the electron ?
    B. What is the kinetic energy of the electron ?
    C. What is the total energy of the electron ?

Chapter 26 Quantum Physics

Chapter Objectives

Upon completion of the unit on quantum physics, students should be able to :

1. state and apply Wien's displacement law.
2. describe the effect of the ultraviolet catastrophe.
3. apply Planck's hypothesis to determine the energy of a photon.
4. describe and/or define the following terms: photoelectric effect, cut-off frequency, stopping potential, threshold wavelength, and work function.
5. identify graphs explaining properties for the photoelectric effect.
6. state and apply the expression for the photoelectric effect.
7. explain how photoelectric effects differ from properties predicted by classical theory.
8. describe the Compton experiment and explain its importance to quantum effects.
9. calculate the Compton wavelength.
10. explain the dual nature of light.
11. describe the Bohr theory of the hydrogen atom.
12. write expressions for finding the energy levels for a hydrogen atom, and construct an energy level diagram for a hydrogen atom.
13. calculate the wavelength, frequency, and energy of photons emitted between levels for an energy diagram.
14. write an expression for the principle quantum number.
15. explain how light is produced in a laser, and give characteristics of the light produced by a laser.

Chapter Summary

*__Thermal radiation__ is the spectrum of radiation emitted by a hot object. A black body is an ideal system that absorbs (and emits) all radiations incident on it. The temperature dependence of the radiation component of maximum intensity is given by **Wein's displacement law**, and λ_{max} is proportional to 1 / T, where T is the absolute temperature. Classically, the intensity of a black body radiation is predicted to be proportional to $1/\lambda^4$, which leads to what is sometimes called the **ultraviolet catastrophe**-- "ultraviolet" because the disagreement with the experiment occurs for short wavelengths beyond the violet end of the visible spectrum, and "catastrophe" because it predicts that the emitted intensity or energy will be infinitely large.

*According to **Planck's hypothesis**, the energy of a thermal oscillator is quantized, that is, an oscillator can have only discrete amounts of energy. The smallest possible amount of oscillator energy is given by E = hf, where h is Planck's constant ($h = 6.63 \times 10^{-34}$ J-s). This is called a **quantum** of energy. Planck's hypothesis gave theoretical predictions that agreed with experimental thermal radiation data.

*A quantum or packet of light is referred to an a **photon**, and in this context, light is transmitted as discrete quanta or "particles" of energy with the energy of a quantum given by E = hf. Einstein used this quantum concept to explain the **photoelectric effect**, which is described mathematically by $hf = K_{max} + \phi_o$. ϕ is the **work function** and $\phi_o = hf_o$, where f_o is the **cut-off frequency**.

*The **Compton effect** is a shift in the wavelength of X-rays scattered by a material. This is explained by treating the X-rays as being composed of quanta or "particles" of energy in elastic collisions (conservation of momentum and kinetic energy).

*The wave theory and the quantum theory of light gave rise to a description called the **dual nature of light**. That is, light sometimes apparently behaves as a wave and at other times as photons or "particles".

*The **Bohr theory of the hydrogen atom** explained the discrete line spectrum of hydrogen. The quantization of the hydrogen electron's angular momentum was introduced into a classical framework, which gave rise to the quantization of orbital radii and energy levels. In making a transition between energy levels, discrete amounts of energy are emitted. The frequencies of these quanta correspond directly to those of the observed spectral lines.

*Quantum theory gave rise to the development of the **laser**, which is an acronym for light amplification by stimulated emission of radiation. Stimulated emission is the process in which a photon with an energy equal to an allowed transition strikes an atom in an excited state, stimulating the atom to make a transition and emit a photon, This process allows the amplification of light -- one photon in, two photons out. Lasing requires having more atoms in an excited state than in the ground state or a population inversion. This can be obtained by some "pumping" or energy input process. Laser beams are intense, highly directional, coherent, and monochromatic.

Chapter 26

Important Terms and Relationships

26.1 Quantization : Planck's Hypothesis

thermal radiation
blackbody
Wein's displacement law : $\lambda_{max}T = 2.90 \times 10^{-3}$ m-K
Planck's constant
Planck's hypothesis

26.2 Quanta of Light : Photons and the Photoelectric Effect

quantum
photon
photon energy : $E = hf$
photoelectric effect
stopping potential
work function
threshold frequency
K_{max} and stopping potential in the photoelectric effect : $K = eV_o$
energy conservation in the photoelectric effect : $hf = K + \phi$
energy conservation (with work function) in photoelectric effect : $hf = K_{max} + \phi_o$
work function and threshold frequency in the photoelectric effect : $\phi_o = hf_o$

26.3 Quantum "Particles" : The Compton Effect

Compton effect : $\Delta\lambda = \lambda_1 - \lambda_o = \lambda_c\,(1 - \cos\theta)$
quantum momentum : $p = h/\lambda$
dual nature of light

26.4 The Bohr Theory of the Hydrogen Atom

emission spectrum
absorption spectrum
Balmer series
Bohr theory of the hydrogen atom
principal quantum number

ground state
excited states
binding energy
lifetime

Bohr theory orbit radius : $r_n = (0.53)n^2 Å = (0.053)n^2$ nm

Bohr theory electron energy : $E_n = -13.6 \text{ eV} / n^2 \quad n = 1,2,3,...$

Bohr theory transition energy (in eV) : $\Delta E = (13.6 \text{ eV}) [(1/n_f^2) - (1/n_i^2)]$

Bohr theory photon wavelength : $\lambda = 12{,}400 \text{ Å} / \Delta E = 1{,}240 \text{ nm}/\Delta E$

26.5 A Quantum Success : The Laser

laser
phosphorescence
metastable state
stimulated state
stimulated emission
population inversion
holography

Additional Solved Problems

26.1 Quantization : Planck's Hypothesis

Example A star gives off light with a frequency of 7.5×10^{14} Hz. Find the approximate temperature of the star.

given : $f = 7.5 \times 10^{14}$ Hz ; $c = \lambda f$; $(3.0 \times 10^8 \text{ m/s}) = \lambda (7.5 \times 10^{14} \text{ Hz})$

$\lambda = 4.0 \times 10^{-7}$ m

$\lambda_{max} T = (2.90 \times 10^{-3} \text{ m-K})$

$(4.0 \times 10^{-7} \text{ m}) T = (2.90 \times 10^{-3} \text{ m-K})$; $T = 7.3 \times 10^3$ K

26.2 Quanta of Light : Photons and the Photoelectric Effect

Example 1 A photon has a wavelength of 500 nm. What is the energy of the photon ?

given : $\lambda = 500$ nm $= 5.00 \times 10^{-7}$ m

first find the frequency : $c = \lambda f$; $(3.0 \times 10^{8}$ m/s$) = (5.00 \times 10^{-7}$ m$)$ f

$$f = 6.0 \times 10^{14} \text{ Hz}$$

$E = hf$; $E = (4.14 \times 10^{-15}$ eV-s$)$ $(6.0 \times 10^{14}$ Hz$) = 2.5$ eV

Example 2 The threshold wavelength for a metal surface is 350 nm. Light strikes the surface with a wavelength of 200 nm.

A. What is the cut-off frequency for the metal ?

B. What is the kinetic energy of the emitted photoelectrons ?

C. What is the stopping potential ?

given : $\lambda_o = 350$ nm ; $\lambda = 200$ nm

(a) $c = \lambda f$; $(3.0 \times 10^{8}$ m/s$) = (350 \times 10^{-9})$ f_o ; $f_o = 8.6 \times 10^{14}$ Hz

(b) first calculate the work function

$\phi = hf_o$; $\phi = (4.14 \times 10^{-15}$ eV-s$)(8.6 \times 10^{14}$ Hz$) = 3.5$ eV

then find the kinetic energy :

$hf = K_{max} + \phi$

$(4.14 \times 10^{-15}$ eV-s$)[\,(3.0 \times 10^{8}$ m/s$)$ / $(200 \times 10^{-9}$ m$)] = K_{max} + 3.5$ eV

6.2 eV $= K_{max} + 3.5$ eV

$K_{max} = 2.7$ eV

Example 3 A metal surface has a work function of 4.5 eV. Light strikes the metal with the following wavelengths. Find the kinetic energy of the photoelectrons.

A. 500 nm.

B. 200 nm.

given : $\phi = 4.5$ eV ; $\lambda_1 = 500$ nm ; $\lambda_2 = 200$ nm ; $h = 4.14 \times 10^{-15}$ eV-s

(a) $c = \lambda f$; 3.0×10^{8} m/s = $(500 \times 10^{-9}$ m$)\, f_1$; $f_1 = 6.0 \times 10^{14}$ Hz

$hf = K_{max} + \phi$

$(4.14 \times 10^{-15}$ eV-s$)(6.0 \times 10^{14}$ Hz$) = K_{max} + 4.5$ eV

2.5 eV $= K_{max} + 4.5$ eV

$K_{max} = -2.0$ eV ; since K cannot be negative, there is no photoelectric emission.

(b) $c = \lambda f$; $(3.0 \times 10^{8}$ m/s$) = (200 \times 10^{-9}$ m$)\, f_2$; $f_2 = 1.5 \times 10^{15}$ Hz

$hf = K_{max} + \phi$

$(4.14 \times 10^{-15}$ eV-s$)(1.5 \times 10^{15}$ Hz$) = K_{max} + 4.5$ eV

6.2 eV $= K_{max} + 4.5$

$K_{max} = 1.7$ eV

26.3 Quantum "Particles" : The Compton Effect

Example A X-ray photon has a wavelength of 2.0×10^{-10} m is scattered through a metal surface. When scattered, the wavelength shift is 2.0×10^{-12} m. What is the scattering angle ?

given : $\lambda_c = 2.4 \times 10^{-12}$ m (constant) ; $\Delta\lambda = 2.0 \times 10^{-12}$ m

$\Delta\lambda = \lambda_c\,(1 - \cos\theta)$

$(2.0 \times 10^{-12}$ m$) = (2.4 \times 10^{-12}$ m$)\,(1 - \cos\theta)$

$0.83 = 1 - \cos\theta$; $\cos\theta = 0.17$; $\theta = 80°$

26.4 The Bohr Theory of the Hydrogen Atom

Example 1 In a hypothetical atom (not hydrogen), an electron falls from an excited energy level of -1.0 eV to an excited state, energy -4.0 eV, and then to the ground state whose energy level is -8.0 eV.

(a) What is the energy of the photon in each transition ?
(b) What is the wavelength of the photon in each transition ?
(c) Are any of these photons visible ?

(a) $E_1 = (-1.0 \text{ eV}) - (-4.0 \text{ eV}) = 3.0 \text{ eV}$
$E_2 = (-4.0 \text{ eV}) - (-8.0 \text{ eV}) = 4.0 \text{ eV}$
(b) $E = hf$
$3.0 \text{ eV} = (4.14 \times 10^{-15} \text{ eV-s}) f$; $f_1 = 7.2 \times 10^{14} \text{ Hz}$
$4.0 \text{ eV} = (4.14 \times 10^{-15} \text{ eV-s}) f$; $f_2 = 9.7 \times 10^{14} \text{ Hz}$
$c = \lambda f$; $(3.0 \times 10^8 \text{ m/s}) = \lambda_1 (7.2 \times 10^{14} \text{ Hz})$; $\lambda_1 = 4.2 \times 10^{-7} \text{ m}$
$(3.0 \times 10^8 \text{ m/s}) = \lambda_2 (9.7 \times 10^{14} \text{ Hz})$; $\lambda_2 = 3.1 \times 10^{-7} \text{ m}$
(c) λ_1 is visible since it is the visible range of light 400 nm to 700 nm
λ_2 is ultraviolet light since its wavelength is slightly less than visible light.

Example 2 An hydrogen electron moves from the n_4 level to the n_2 level.
A. Find the energy of the emitted photon.
B. Find the wavelength of the emitted photon ?

$E = -13.6 \text{ eV} / n^2$
$E_4 = -13.6 \text{ eV} / 4^2 = -0.85 \text{ eV}$; $E_2 = -13.6 \text{ eV} / 2^2 = -3.40 \text{ eV}$

A. $E_{photon} = E_4 - E_2 = (-0.85 \text{ eV}) - (-3.40 \text{ eV}) = 2.55 \text{ eV}$
B. $E = hf$
$2.55 \text{ eV} = (4.14 \times 10^{-15} \text{ eV-s}) f$; $f = 6.16 \times 10^{14} \text{ Hz}$
$c = \lambda f$; $(3.0 \times 10^8 \text{ m/s}) = \lambda (6.16 \times 10^{14} \text{ Hz})$; $\lambda = 4.87 \times 10^{-7}$ m or 487 nm

Solutions to paired problems and other selected problems

6. $\lambda_m = (2.90 \times 10^{-3}) / T$
$450 \times 10^{-9} \text{ m} = (2.90 \times 10^{-3}) / T$; $T = 6.44 \times 10^3 \text{ K}$

11. $E = hf$; $c = \lambda f$; $E = hc / \lambda$; $E = hc\, T / (2.90 \times 10^{-3})$
$E = (6.63 \times 10^{-34} \text{ J-s})(3.0 \times 10^8 \text{ m/s})(373 \text{ K}) / (2.90 \times 10^{-3} \text{ K}) = 2.56 \times 10^{-20} \text{ J}$

14. (d)

22. $E = hc/\lambda$; $E = (6.63 \times 10^{-34}\ \text{J-s})(3.0 \times 10^{8}\ \text{m/s}) / (550 \times 10^{-9}\ \text{m}) = 3.6 \times 10^{-19}\ \text{J}$
$E = Pt$; $E = (5.0\ \text{W})(60\ \text{s}) = 300\ \text{J}$
$n = (300\ \text{J}) / (3.6 \times 10^{-19}\ \text{J}) = 8.3 \times 10^{20}$ quanta

30. (a) $K = hf - \phi$; $c = \lambda f$; $(3.00 \times 10^{8}\ \text{m/s}) = (160 \times 10^{-9}\ \text{m})f$; $f = 1.88 \times 10^{15}\ \text{Hz}$
$K = (6.63 \times 10^{-34}\ \text{J-s})(1.88 \times 10^{15}\ \text{Hz}) - (7.71 \times 10^{-19}\ \text{J}) = 4.69 \times 10^{-19}\ \text{J}$
(b) $\phi = hf_o$; $(7.71 \times 10^{-19}\ \text{J-s}) = (6.63 \times 10^{-34}\ \text{J-s})f_o$; $f_o = 1.16 \times 10^{15}\ \text{Hz}$

40. $\Delta\lambda = \lambda_c(1 - \cos\theta)$; $7.11 \times 10^{-13}\ \text{m} = (2.43 \times 10^{-12}\ \text{m})(1 - \cos\theta)$
$\theta = 45°$

44. (d)

52. $E_n = 13.6 / n^2$
(a) $E_3 = 13.6\ \text{eV} / 3^2 = 1.51\ \text{eV}$
(b) $E_6 = 13.6\ \text{eV} / 6^2 = 0.38\ \text{eV}$
(c) $E_{10} = 13.6\ \text{eV} / 10^2 = 0.14\ \text{eV}$

56. (a) $\Delta E_{52} = (13.6\ \text{eV})[(1/2^2) - (1/5^2)] = 2.86\ \text{eV}$
$\Delta E_{21} = (13.6\ \text{eV})[(1/2^2) - (1/1^2)] = 10.2\ \text{eV}$
$\lambda_{52} = 1240\ \text{nm} / \Delta E = 1240\ \text{nm} / 2.86 = 434\ \text{nm}$
$\lambda_{21} = 1240\ \text{nm} / 10.2 = 122\ \text{nm}$
(b) the transition from 5 to 2 is in the visible range ; the transition 2 to 1 is in the ultraviolet region.

60. $E = -ke^2 / 2r$
$E = -(9.0 \times 10^{9}\ \text{N-m}^2/\text{C}^2)(1.6 \times 10^{-19}\ \text{C})^2 / [2(5.29 \times 10^{-11}\ \text{m})]$
$E = -2.18 \times 10^{-18}\ \text{J}$ or $-13.6\ \text{eV}$

66. $E^2 = p^2c^2 + (m_oc^2)^2$; $m_o = 0$
$E = p^2c^2$; $p = E/c = hf/c = h/\lambda$

Chapter Quiz

Multiple Choice. Choose the best answer.

__ 1. Light is detected from a star. Which star would have the highest temperature ?
A. red B. blue C. yellow D. ultraviolet

__ 2. If the wavelength of a photon is doubled, by what factor does the energy change ?
A. 1/4 B. 1/2 C. 2 D. 4

__ 3. The intensity of light incident on a photosensitive material is doubled. By what factor does the kinetic energy of a photoelectron change ?
A. 1/2 B. 1 C. 2 D. 4

__ 4. A hypothetical atom has three excited states in addition to its ground state. How many different spectral lines are possible ?
A. 3 B. 4 C. 5 D. 6

__ 5. In a photoelectric experiment, the threshold wavelength of a metal is 500 nm. Light incident on the surface has a wavelength of 600 nm. Which of the following describes what takes place ?
A. a photoelectric emission occurs with positive kinetic energy.
B. a photoelectric emission occurs with negative kinetic energy.
C. the electron is broken away but has no kinetic energy.
D. there is no photoelectric emission.

__ 6. A surface cutoff frequency of 4.0×10^{14} Hz. What is the work function ?
A. 2.7×10^{-19} eV B. 1.7 eV C. 5.4×10^{-19} J D. 3.4 eV

__ 7. A helium atom is four times more massive as a hydrogen atom. Compared to the size of a hydrogen atom, a helium atom is
A. four times larger B. twice as large C. the same size D. smaller

Problems

8. Find the wavelength and the frequency of the light from a body whose temperature is 2000 K.

9. The cut-off frequency for a surface is 2.0×10^{14} Hz. Light with a wavelength of 500 nm is incident on the surface.
   A. Find the threshold wavelength.
   B. Find the kinetic energy of the emitted electron.
   C. Find the stopping potential.

10. An hydrogen electron moves from the n = 6 level to the n = 2 level.
    A. What is the frequency and wavelength of the light ?
    B. Is the light visible ?

CHAPTER 27 Quantum Mechanics

Chapter Objectives

Upon completion of the unit on quantum mechanics, students should be able to :

1. apply the deBroglie hypothesis, and calculate the wavelength of a particle.
2. describe the purpose and importance of the Davisson-Germer experiment to the understanding of the wave nature of electrons.
3. apply the wave properties of electrons to determine the diffraction pattern for an electron.
4. apply the Schrodinger wave equation to determine the behavior of particles in terms of probability.
5. state and apply the Heisenburg uncertainty principle.
6. compare and contrast matter and antimatter, and describe the interaction of the two.

Chapter Summary

*The waves associated with moving particles are called **matter waves** or **de Broglie waves**. The Davisson-Germer experiment experimentally demonstrated the wave-like properties of particles.

*The **Schrodinger wave equation**, describes the de Broglie matter waves $(K + U)\Psi = \Psi E$. The term Ψ term is called the **wave function and** describes the wave as a function of time and space. Ψ^2 represents the possibility of finding the particle at a certain position and time, or probability density. In some instances, a a particle has a finite probability of being in a classically forbidden region.

*According to the **Heisenburg uncertainty principle**, as applied to position and momentum (or velocity), it is impossible simultaneously to know the object's exact position and momentum $(\Delta p)(\Delta x) \geq h/2\pi$. The uncertainties in energy and time have a similar relationship, $(\Delta E)(\Delta t) \geq h/2\pi$.

*The positron, which has the same mass as an electron but a positive electronic charge, is the antiparticle of the electron. A positron can only be created with the simultaneous creation of an electron in a process called **pair production**. Positrons and electrons are "destroyed" or converted into energy in a process called **pair annihilation**. All subatomic particles have been found to have antiparticles. Atoms of **antimatter** consist of negatively charged nuclei composed of antiprotons and antineutrons, surrounded by positively charged positrons (antielectrons).

Important Terms and Relationships

27.1 Matter Waves : The deBroglie Hypothesis

quantum mechanics
deBroglie hypothesis
deBroglie (matter) waves
momentum of a photon : $p = E / c = hf / c = h / \lambda$
deBroglie wavelength : $\lambda = h / p = h / mv$
electron wavelength when accelerated through potential V :
$\lambda = (150 / V)^{1/2}$ Å

27.2 The Schrodinger Equation

wave function Ψ
Schrodinger's wave equation
probability density (Ψ^2)

27.3 Atomic Quantum Numbers and the Periodic Table

orbital quantum number (l)
magnetic quantum number (m_l)
spin quantum number (m_s)
shell
subshell
Pauli exclusion principle
electron configuration
electron period
periodic table of elements
period
group

27.4 The Heisenburg Uncertainty Principle

$(\Delta p)(\Delta x) \geq h / 2\pi$
$(\Delta E)(\Delta t) \geq h / 2\pi$

27.5 Particles and Antiparticles

positron
antiparticle
pair production
pair annihilation
antimatter
condition for electron pair production : $hf \geq 2m_ec^2 = 1.022$ MeV

Additional Solved Problems

27.1 Matter Waves : The deBroglie Hypothesis

Example 1 Deterine the deBroglie wavelength of a 1200 kg car moving with a speed of 22 m/s.

$\lambda = h / m_o v$

$\lambda = (6.63 \times 10^{-34}$ J-s) / [(1200 kg)(22 m/s)] ; $\lambda = 2.5 \times 10^{-38}$ m

Example 2 Determine the deBroglie wavelength of an electron moving with a speed of 0.90 c.

$\lambda = h / mv$
the relativistis mass must be found :

$\gamma = 1 / [1 - v^2 / c^2]^{1/2}$; $\gamma = 1 / [1 - (0.90\, c / c)^2]^{1/2} = 2.3$

$m = \gamma m_o = (2.3)(9.1 \times 10^{-31}$ kg) $= 2.1 \times 10^{-30}$ kg

$\lambda = h / mv = (6.63 \times 10^{-34}$ J-s) / [$(2.1 \times 10^{-30}$ kg)$(2.7 \times 10^{8}$ m/s)]

$\lambda = 1.2 \times 10^{-12}$ m

27.4 The Heisenburg Uncertainty Principle

Example 1 An electron moves with a speed of 3.0×10^6 m/s ± 10%. What is the minimum uncertainty in its position ?

$\Delta p = mV_{max} - mv_{min} = (9.1 \times 10^{-31}$ kg) [$(3.3 \times 10^6$ m/s) - $(2.7 \times 10^6$ m/s)]

$\Delta p = 5.46 \times 10^{-25}$ kg-m/s

$\Delta p \, \Delta x \geq h / 2\pi$

$(5.46 \times 10^{-25}$ kg-m/s) $\Delta x \geq (6.63 \times 10^{-34}$ J-s) / 2π

$\Delta x = 1.9 \times 10^{-10}$ m

Example 2 The speed of 500-eV electron is known to be ±10%. How accurately can its position be measured ?

500 eV $(1.6 \times 10^{-19}$ J / eV) = 8.0×10^{-17} J

$K = (1/2)mv^2$

8.0×10^{-17} J = $(1/2)(9.1 \times 10^{-31}$kg) v^2 ; $v = 1.3 \times 10^7$ m/s

$\Delta p = m\,(v_{max} - v_{min}) = (9.1 \times 10^{-31}$ kg)[$(1.4 \times 10^7$ m/s) - $(1.2 \times 10^7$ m/s)]

$\Delta p = 1.8 \times 10^{-24}$ kg-m/s

$\Delta p \, \Delta x \geq h / 2\pi$

$(1.8 \times 10^{-24}$ kg-m/s) $\Delta x \geq (6.63 \times 10^{-34}$ kg-m/s) / 2π

$\Delta x = 5.9 \times 10^{-11}$ m

Solutions to paried problems and other selected problems

4. $\lambda = h / mv$

(a) electron : $\lambda = (6.63 \times 10^{-34}$ J-s) / [$(9.11 \times 10^{-31}$ kg)(300 m/s)] $= 2.43 \times 10^{-6}$ m

proton : $\lambda = (6.63 \times 10^{-34}$ J-s) / [$(1.67 \times 10^{-27}$ kg)(300 m/s)] $= 1.32 \times 10^{-9}$ m

15. $V = (150) / \lambda^2$
$V = (150 \text{ eV}) / (1.0 \text{ Å})^2 = 150 \text{ V}$

18. $n\lambda/2 = L$ or $\lambda = 2L/n$ where $n = 1,2,3,4,...$
satisfied for $\Psi_n = A \sin n\pi x / L$, $n = 1,2,3,4,...$
$p = h/\lambda = nh/2L$
$K = E = p^2/2m = n^2h^2/8mL^2$

22. (a)

30. (a) Be (b) N (c) Ne (d) S

36. $\Delta v = h / 2\pi m \Delta x$; $\Delta v = (6.63 \times 10^{-34} \text{ J-s}) / [2\pi (0.50 \text{ kg})(10^{-5} \text{ m})]$
$\Delta v = 2.1 \times 10^{-29} \text{ m/s}$

44. (d)

47. $E = hf$; $8.2 \times 10^{-14} \text{ J} = (6.63 \times 10^{-34} \text{ J-s}) f$; $f = 1.2 \times 10^{20} \text{ Hz}$

54. function has the required nodal end points

Sample Quiz

Multiple Choice. Choose the best answer.

__ 1. Which of the following would tend to have the greatest deBroglie wavelength (assuming they could travel at the same speed) ?
A. proton B. electron C. baseball D. car

__ 2. An electron is subjected to an electric potential of 150 V. What is the deBroglie wavelength ?
A. 1 m B. 1 cm C. 1 nm D. 1 Å

__ 3. If the wave function were doubled, by what factor does the probability of finding the particle at a certain position and time change ?
A. 1/4 B. 1/2 C. 2 D. 4

__ 4. What is the maximum number of electrons that can occupy the 3rd subshell ?
A. 10 B. 2 C. 6 D. 14

__ 5. What is the electron configuration for Li ?
A. $1s^2$ B. $1s^2 2s^1$ C. $1s^2 2s^2$ D. $1s^1 2s^3$

__ 6. What is the electron configuration for Na ?
A. $1s^2 2s^2 2p^6$ B. $1s^2 2s^2 2p^6 3s^1$ C. $1s^2 2s^2 2p^6 3s^2$ D. $1s^2 2s^2 2p^6 3s^3$

__ 7. As the speed of a particle increases, the uncertainty of its position
A. increases B. decreases C. remains the same

Chapter 27

Problems

8. Write the electron configuration fo N.

9. An electron initially at rest accelerates through an electric potential of 500 V. What is the deBroglie wavelength ?

10. An electron in an excited state has a lifetime of 10^{-9} s. What is the minumum uncertainty in the energy of the photons emitted on de-excitation ?

CHAPTER 28 The Nucleus

Chapter Objectives

Upon completion of the unit on the nucleus and radioactivity, students should be able to :

1. explain the modern concept of the atom, pointing out significant features including charge, motion, force, and energy.
2. name the significant contributions of Rutherford and Thomson in atomic theory.
3. define key terms, including the atomic number, nucleon, mass number, isotope, and nuclide
4. determine the constituent parts of an atom.
5. calculate the binding energy, and binding energy per nucleon for a nucleus.
6. describe radioactivity, and list and compare the three tyes of decay in terms of penetrating ability, mass, and charge.
7. write nuclear transfomations for the three types of decay.
8. write an expression for finding an isotope with a given half-life, the fraction of its nuclei which have decayed as a function of time.
9. determine the half-life of a substance from a graph of the amount of radioactive sample remaining as a function of time.
10. determine the decay rate of a substance.
11. give the different units for measuring radiation.

Chapter Summary

* The simplistic solar-system model of the atom pictures all positive charge concentrated in a very small central region or nucleus surrounded by orbiting electrons. This is sometimes called the Rutherford-Bohr model. Using scattering experiments, the upper limit for the nuclear radius as determined by the distance of closest approach is on the order of 10^{-12} cm.

* The strong **nuclear force** in a nucleus is (1) strongly attractive, and much larger in relative magnitude than the electrostatic or gravitational forces, (2) is very short ranged, (3) acts between any two nucleon within a short range, i.e., between two protons, a proton and an a neutron, or two neutrons.

* In nuclear notation, A is the **mass number** (p+n), Z is the atomic or **proton number** (p), and N is the **neutron number** (n). In symbol form ${}_{Z}^{A}X_{N}$, where X is

the chemical symbol of the element of the nucleus. Isotopes are nuclei with the same number of protons but different numbers of neutrons, A particular nuclear species or isotope is called a **nuclide**.

*Some nuclei decay spontaneously of their own accord and are said to be **radioactive**. Three decay modes are:

(1) Alpha decay, in which the nucleus emits an **alpha particle**, a doubly charged (2+) ion containing two protons and two neutrons, that is identical to the nucleus of a helium atom $(^{4}_{2}He)$.

(2) Beta decay, in which the nucleus emits a **beta particle** or an electron (^{-1}e).

(3) Gamma decay, in which an excited nucleus emits a **gamma ray**, or a "particle" or quantum of electromagnetic energy.

Two laws apply to all nuclear processes: <u>the conservation of nucleons</u> and <u>the conservation of charge</u>.

*Alpha decay involves the quantum mechanical process of "tunneling" or <u>barrier penetration</u>. There are actually three modes of beta decay: $\beta-$, $\beta+$, and electron capture. In gamma decay, the mass and the proton numbers do not change in the process.

*The **activity** of a radioactive isotope is expressed in terms of the number of disintegrations or decays (ΔN) per unit time (Δt), i.e. $\Delta N / \Delta t$, and $\Delta N / \Delta t = -\lambda N$, where N is the number of nuclei present and λ is the **decay constant**, which is different for different isotopes. The **half-life** ($t_{1/2}$) is defined as the time it takes for half of the original nuclei in a radioactive sample to decay. In terms of the decay constant, $t_{1/2} = 0.693 / \lambda$. The SI unit of radioactivity is the **becquerel** (Bq), and 1 Bq = 1 decay/s. Another common unit is the **curie** (Ci), and $1\ Ci = 3.70 \times 10^{10}$ decays/s $= 3.70 \times 10^{10}$ Bq.

*General criteria for nuclear stability are :

(1) All isotopes with proton numbers greater the 83 ($Z > 83$) are unstable.

(2) Most even-even nuclei are stable.
Most odd-even or even-odd nuclei are stable. Most odd-odd nuclei are unstable. ^{2}H, ^{6}Li, ^{10}Be and ^{14}N are stable.

(3) Stable nuclei with A , 40 have approximately the same number of protons and neutrons. Stable nuclei with $A > 40$ have more neutrons that protons.

*The masses of nuclei and nuclear particles are commonly measured in the atomic mass unit (u), 1 u = 1.6606 x 10^{-27} kg, which is referenced to a neutral atom of ^{12}C.

This atom is taken to have an exact mass of 12.000000 u. The **total binding energy** E_b of a nucleus is the energy/mass difference of the total mass of the individual nucleons and the mass of the nucleus. The **average binding energy per nucleon** is E_b/A and gives an indication of nuclear stability -- the greater E_b/A, the more tightly bound are the nucleon in a nucleus.

*Methods of radiation detection include : the Geiger counter, the scintillation counter, the solid state of semiconductor detector, the bubble chamber, and the spark chamber.

*Carbon-14 dating involves measuring the activity of ^{14}C beta decay in once-living things and using the half-life of this isotope to calculate when the activity of the objects was constant and alive.

Important Terms and Relationships

28.1 Nuclear Structure and the Nuclear Force

Rutherford-Bohr model
strong nuclear force
nucleon
proton (atomic) number
mass number
neutron number
isotope
nuclide
radius of nucleus (upper limit) $r = 4kZe^2 / mv^2$

28.2 Radioactivity

radioactivity
alpha particle
beta particle
gamma ray
alpha decay
conservation of nucleons
conservation of charge
tunneling

barrier penetration
beta (β) decay
β^+ decay
electron capture
gamma decay

28.3 Decay Rate and Half-life

activity
activity of a radioisotope : $\Delta N/\Delta t = -\lambda N$
decay constant
half-life
curie (Ci)
becquerel (Bq)
carbon-14 dating

Number of undecayed nuclei : $N = N_o e^{-\lambda t}$

half-life and decay constant : $t^{1/2} = (0.693) / \lambda$

28.4 Nuclear Stability and Binding Energy

pairing effect
atomic mass unit (u)
total binding energy $E_b = (\Delta m)c^2$
average binding energy per nucleon E_b / A

28.5 Radiation Detection and Application

radiation detectors
roentgen (R)
rad
gray (Gy)
rem
relative biological effectiveness (RBE)
sievert (Sv)
neutron activation analysis

effective dose : dose (in rem) = dose (in rad) x RBE
dose (in Sv) = dose (in Gy) x RBE

Additional Solved Problems

28.1 Nuclear Structure and the Nuclear Force

Example In a ^{14}C ion with a charge of +2 ,
A. how many protons are present ?
B. how many neutrons are present ?
C. how many electrons are present ?

(a) the atomic number equals the number of protons 6
(b) A = Z + N
14 = 6 + N ; N = 8 -- 8 neutrons
(c) since there is a charge of +2, this means there are two more protons than electrons : 4.

28.2 Radioactivity

Example Write a nuclear equation for the following.
A. ^{98}Tc undergoes a beta decay.
B. ^{210}Pb undergoes an alpha decay.
C. ^{218}At undergoes an alpha decay.
D. ^{218}At undergoes a beat decay.
E. ^{239}Pu undergoes a gamma decay.

A. $^{98}_{43}Tc \Rightarrow {}^{98}_{44}Ru + {}^{0}_{-1}e$

B. $^{210}_{82}Pb \Rightarrow {}^{206}_{80}Hg + {}^{4}_{2}He$

C. $^{218}_{85}At \Rightarrow {}^{214}_{83}Bi + {}^{4}_{2}He$

D. $^{218}_{85}At \Rightarrow {}^{218}_{86}Rn + {}^{0}_{-1}e$

E. $^{239}_{94}Pu \Rightarrow ^{239}_{94}Pu + \gamma$

28.3 Decay Rate and Half-life

Example 1 The half-life for a certain material is 5.0 min. What fraction of the material is left after 30 min ?

30 min / 5.0 min = 6 ; $1 / 2^n = 1 / 2^6$ = (1 / 64) of the original material

Example 2 The half-life of a 2.0 kg material is 10 min. How much of the material is left after 25 min ?

$N = N_o e^{-\lambda t}$; λ =(0.693) / 10 min

$N = (2.0\text{ kg})\,(2.718)^{-[(0.693)\,(25\text{ min})\,/\,(10\text{ min})]}$

N = 0.35 kg

28.4 Nuclear Stability and Binding Energy

Example 1 The mass of ^{132}Xe is 131.90415 u. What is the binding energy per nucleon for this isotope ?

Xe-132

54 (^1H) = 54 (1.007825 u) = 54.42255 u
78 (^1n) = 78 (1.008665 u) = 78.67587 u
total 133.09842 u
- mass Xe-132 131.90415 u
1.19427 u (x 931.5 MeV / u) = 1112 MeV

1112 / 132 = 8.43 MeV / nucleon

Example 2 What is the binding energy for a ^7Li atom. The mass of ^7Li is 7.016005 u.

Li - 7 has 3 protons and 4 neutrons

$3\,(^1H) = 3\,(1.007825\ u)$

$4\,(^1n) = 4\,(1.008665\ u)$ total mass 7.058135 u

\- mass ^7Li $\underline{7.016005\ u}$

$4.213 \times 10^{-2}\ u\ (931.5) = 39.2\ MeV$

28.5 Radiation Detection and Application

Example A radioactive isotope of ^{131}I is 8.0 days. The initial mass is 2.0 g. How many iodine atoms decay per second ?

$N_0 = 3.5\ g\ (1\ mole / 131\ g)\ (6.02 \times 10^{23}\ atoms/mole) = 1.6 \times 10^{22}\ atoms$

$T_{1/2} = 8.0\ days\ (24\ h / day)\ (3600\ s / h) = 6.9 \times 10^5\ s$

$\lambda = (0.693) / T_{1/2}\ ;\ \lambda = (0.693) / (6.9 \times 10^5\ s) = 1.0 \times 10^{-6}\ s^{-1}$

$\Delta N/\Delta t = -\lambda N = -(1.0 \times 10^{-6}\ s^{-1})(1.6 \times 10^{22}\ atoms) = 1.6 \times 10^{16}\ atoms/s$

Solutions to paired problems and additional solved problems

2. (d)

8. $^{16}_{8}O$; $^{17}_{8}O$; $^{18}_{8}O$

18. (a) $^{222}_{86}Rn \Rightarrow {}^{218}_{84}Po + {}^{4}_{2}He$

(b) $^{218}_{84}Po \Rightarrow {}^{214}_{82}Pb + {}^{4}_{2}He$

$^{218}_{84}Po \Rightarrow {}^{218}_{85}At + {}^{0}_{-1}e$

30. (a) 1/4

(b) $1 / 2^n = 1/2^{24}$ or 6.0×10^{-8}

36. $1\,t_{1/2}$: $N / N_o = 1/2$

$2\,t_{1/2}$: $N / N_o = 1/4$

$3\,t_{1/2}$: $N / N_o = 1/8$

therefore : $N / N_o = 1 / 2^n$

42. $\lambda = (0.693) / t_{1/2}$; $\lambda = (0.693) / 21.8 = 0.0318\ \text{min}^{-1}$

moles = $(25 \times 10^{-3}) / 223 = 1.12 \times 10^{-4}$

(a) $N_o = nN_a = (1.12 \times 10^{-4})(6.02 \times 10^{23}) = 6.74 \times 10^{19}$ nuclei

(b) $N = N_o\, e^{-t\lambda} = 6.74 \times 10^{19})\, e^{-(109)(0.0318)} = (6.74 \times 10^{19})\, e^{-3.47} =$

2.11×10^{18} nuclei

48. (d)

56. 4 ^{1}H : 4 (1.007825 u)

5 ^{1}n : 4 (1.008665 u)

sum of the parts : 8.065960 u

(8.065960 u) - (9.012183 u) = 0.062442 u ;

E = (0.062442 u)(931.5 MeV / u) = 58.2 MeV ; E_b / a = 6.46 Mev / nucleon

60. mass of ^{27}Al = mass of alpha + mass of daugter + Δm

26.981541 u = 4.002603 u + 22.98770 + Δm

Δm = 0.01943 u ; E_b = (0.01943 u)(931.5 MeV / u) = 10.1 MeV

66. (d)

70. dose (in rem) = dose (in rad) x RBE

(0.50)(1) + (0.30)(4) + (0.10)(20) = 3.7 rem

this exceeds the 3.0 rem limit

Sample Quiz

Multiple Choice. Choose the correct answer.

Questions 1 and 2 refer to the following : $^{15}_{8}O$

__ 1. How many protons are there in the atom ?
A. 7 B. 8 C. 15 D. 23

__ 2. How many neutrons are there in the atom ?
A. 7 B. 8 C. 15 D. 23

__ 3. Which type of decay has the most mass ?
A. alpha B. beta C. gamma D. x-ray

__ 4. Which type of decay cannot experience a magnetic force ?
A. alpha B. beta C. gamma D. x-ray

__ 5. The number of protons in an atom is 98 and the mass number is 249. It undergoes a double alpha decay. What is the number of protons and the mass number of the final nucleus ?
A. 90 , 245 B. 94 , 241 C. 100 , 249 D. 100 , 257

__ 6. A substance has a half-life of 15 min. What fraction of the material is left after 1 hour ?
A. 0 B. 1/2 C. 1/8 D. 1/16

__ 7. Which of the following has the greatest binding energy ?
A. helium B. carbon C. mercury D. uranium

Problems

8. Write the nuclear equations for the following :

   A. ${}^{145}_{61}Pm$ undergoes an alpha decay.

   B. ${}^{145}_{61}Pm$ undergoes a beat decay

9. Ten grams of a certain isotope has a half-life of 1 hour.

   A. How much of the material is left after 4 h ?

   B. How much of the material is left after 90 minutes ?

10. Determine the binding energy and the binding energy per nucleon for ${}^{191}Ir$. The mass of ${}^{191}Ir$ is 190.96060 u.

CHAPTER 29 Nuclear Reactions and Elementary Particles

Chapter Objectives

Upon completion of the unit on nuclear reactions and elementary particles, students should be able to :

1. write nuclear equations applying the conservation of charge and mass.
2. determine the Q value for a nuclear reaction, and determine of the reaction is exoergic or endoergic.
3. determine the threshold energy for an endoergic reaction.
4. distinguish between fission and fusion.
5. find the energy released in a fission reaction and in a fusion reaction.
6. explain how a breeder reactor differs from a conventional reactor.
7. define and/or explain the following : chain reaction, critical mass, fuel rod, control rods, moderator, and plasma.
8. state the four fundamental forces and the virtual exchange particle for each fundamental force.
9. explain the role of the neutrino in a beta decay.
10. state the different fundamental quantities, and explain the role of each.

Chapter Summary

*Nuclear reactions can be produced by energetic particles from radioactive sources or from particle accelerators. The **Q value** is a measure of the total energy released or absorbed in a reaction. When energy is released ($Q > 0$), the reaction is said to be **exoergic** (or exothermic). When energy is absorbed ($Q < 0$), the reaction is said to be **endoergic** (or endothermic).

*The minimum kinetic energy that an incident particle needs to initiate an endoergic reaction is called the **threshold energy**. When a particle has more kinetic energy than the threshold energies of several possible reactions, any of the reactions may occur. A measure of the probability that a particular reaction will occur is called the **cross section** of the reaction.

*In nuclear **fission**, a heavy nucleus divides into two lighter nuclei with the emission of two or more neutrons. The energy release in a fission reaction is about 1 MeV per nucleon of fissioning nucleus, or on the order of a total of 200 MeV. A sustained release of nuclear energy can be accomplished by a **chain reaction**, in which neutrons from one fission reaction induces other fission reactions. To have a sustained chain reaction, there must be a minimum or **critical mass** of the fissionable material. Basically, enough mass is needed so that neutrons do not escape without causin fission reactions.

*In nuclear **fusion**, light nuclei fuse together to form a heavier nucleus, with the release of energy. The energy release from a D-T reaction is about 17 MeV.

*In a fission reactor, enriched uranium fuel pellets are contained in **fuel rods** in a fuel rod assembly. Water flowing through the fuel rod assembly not only acts as a coolant , but also as a **moderator**. Through collisions with the hydrogen atoms in the water, neutrons are "slowed" or moderated, which increases the reaction cross section for U-235 fission. A nuclear reactor, such as used in electrical generation, cannot explode like a nuclear or "atomic" bomb (uncontrolled fission).

*A breeder reactor produces more fissionable fuel than it consumes. U-235 is consumed and fissionable Pu-239 is produced from neutron reactions with U-238.

* The terms LOCA and meltdown refer to loss of coolant accident, in which the reactor coolant might be lost or stop flowing through the core, causing the reactor to overheat, and the core to meltdown into a hot, fissioning mass that could melt through the floor of the containment building into the environment.

*Fusion by thermonuclear reactions requires high temperatures than ionize the fusion material into a gas of charged particles called a **plasma**. Research is being done on controlled fusion processes. Plasma confinement is a major problem, and the two approaches are magnetic confinement and inertial confinement.

*An additional particle, other than the nuclei and the electron, is needed in beta decay to account for the apparent violations of energy, linear momentum, and spin. This particle is the **neutrino**. The neutron interacts with matter by the weak nuclear force. Neutrinos are produced in β^+ decay reactions and antineutrons are produced in β^-reactions.

*Fundamental forces are considered to be **virtual exchange particles**. The exchange particle for the electromagnetic force is a virtual photon. The short-range strong nuclear force between two nucleons is associated with an exchange particle called the meson, viz., the μ meson or muon. The virtual exchange particles for the weak nuclear force is the W particle (weak). The exchanges particle for the gravitational forces is called the graviton. The existence of gravitons has not been experimentally confirmed.

*__Elementary particles__ are fundamental particles, or building blocks, of atoms. **Hadrons** are particles that interact by the strong nuclear force (e.g., protons , neutrons, and pions). **Leptons** are particles that interact by the weak nuclear force, but not the strong nuclear force (e.g. , electrons, muons, and neutrinos). **Quarks** are elementary particles that make up hadrons. Quarks combine only in two possible ways, either in three's or in quark-antiquark pairs. Three-quark combinations are called baryons, and quark combinations are called mesons.

*There are six **flavors** of types of quarks: up (u), down (d), strange (s), charm (c), top or truth (t), and bottom or beauty (b). Quarks have fractional electronic charges of either (-1/3) or (+2/3). Quarks interact by the strong force, but they are also subject to the weak force. A weak force acting on a quark changes its flavor, and gives rise to the decay of hadrons.

*The exchange particle for quarks is the **gluon**. To give the strong force a field representation, each quark is said to possess an analog of electric charge which is the source of the "gluon field". Instead of charge, this property is called **color** (no relationship to ordinary color). Each quark can come in one of three possible colors: red, blue and green, There are corresponding anticolors for antiquarks. When a quark emits or absorbs a gluon, it changes color.

*The electromagnetic force and the weak force are two parts of a single **electroweak force**. A theory that would merge the strong nuclear force and the electroweak force is called the **grand unified theory** (GUT). Perhaps, all forces are part of a single **superforce**, which is a real theoretical challenge.

Important Terms and Relationships

29.1 Nuclear Reactions

nuclear reactions
particle accelerators
Q value : $Q = (m_A + m_a - m_B - m_b)c^2 = (\Delta m)c^2$
endoergic
exoergic
threshold energy : $K_{min} = [\,1 + (m_a/M_A)]\,|Q|$
cross section

29.2 Nuclear Fission

fission reaction
chain reaction
critical mass
nuclear reactor
fuel rods
control rods
moderator
breeder reactor
LOCA
meltdown

29.3 Nuclear Fusion

fusion (thermonuclear) reaction
plasma
magnetic confinement
inertial confinement

29.4 Beta Decay and the Nutrino

neutrino

29.5 Fundamental Forces and Exchange Particles

fundamental forces
virtual particle
exchange particle
photon
muon
pion
weak nuclear force
W particle
graviton

29.6 Elementary Particles

elementary particles
hadrons
leptons
tauon
quarks
barons
gluon
electroweak force

Additional Solved Problems

29.1 Nuclear Reactions

Example 1 In the following find the threshold energy

$$^{13}C + {}^{1}H \Rightarrow {}^{13}N + {}^{1}n$$

(13.003355 u) (1.007825 u) (13.005739 u) (1.008665 u)

(13.003355 u) + (1.007825 u) = 14.01118 u
(13.005739 u) + (1.008665 u) = 14.01440 u
(14.01118 - 14.01440) (931.5) = 3.00 MeV

Example 2 Is the following exoergic or endoergic ?

$$^{7}Li + {}^{1}p \Rightarrow {}^{4}He + {}^{4}He$$

(7.016005 u) (1.007825 u) (4.002603 u) (4.002603 u)

(7.016005 u) + (1.007825 u) = 8.023830 u
(4.002603 u) + (4.002603 u) = 8.005206 u

Q = (8.023830 u - 8.005206 u) x 931.5 = 17.3 MeV

$K_{min} = (1 + m/M)\,Q$
K_{min} = [1 + (1.007825 u / (7.016005 u)] (17.3 MeV) = 19.8 MeV

29.2 Nuclear Fission

Example Find the energy release in the following fission reaction.

$^{1}n + {}^{235}U \Rightarrow {}^{141}Ba + {}^{92}Kr + 3\,{}^{1}n$

(1.008665 u) (235.043925 u) (140.9142 u) (91.9252 u) (1.008665 u)

(1.008665 u + 235.043925 u) - (140.9142 u + 91.9252 + 3(1.008665)] = 0.187 u
energy = (0.187 u) (931.5 MeV / u) = 174 MeV

29.3 Nuclear Fusion

Example Find the energy release in the following :

$^{2}H + {}^{2}H \Rightarrow {}^{3}He + {}^{1}n$
(2.014102 u) (3.016029 u) (1.008665 u)

(2.014102 u + 2.014102 u) - (3.016029 u + 1.008665 u) = 3.51×10^{-3} u
Energy = (3.51×10^{-3} u) (931.5 MeV / u) = 3.27 MeV

29.4 Beta Decay and the Neutrino

Example Find the energy release in the following decay.

$^{32}_{15}P \Rightarrow {}^{0}_{-1}e + {}^{32}_{16}S$

(31.973908 u) (31.972072 u)

31.973908 u - 31.972072 u = 1.84×10^{-3} u
(1.84×10^{-3} u)(931.5 MeV /u) = 1.71 MeV

Solutions to paired problems and other selected problems

2. (d)

8. $^{211}_{82}Pb \Rightarrow ^{211}_{83}Bi + ^{0}_{-1}e$

   210.98726 u - 210.98726 u = (0.00148 u)

   (0.00148 u)(931.5 MeV / u) = 1.379 MeV

   1.379 MeV - 0.50 MeV = 0.868 MeV

19. $^{3}_{1}H + ^{1}_{1}H \Rightarrow ^{2}_{1}H + ^{2}_{1}H$

   (3.016049 u + 1.007825 u) - 2 (2.01402 u) = -0.00423 u)

   (0.00423 u)(931.5 MeV / u) = 3.94 MeV

   K = [1 + (1.007825 u) / (3.016049 u)] (3.94 MeV) = 5.37 MeV

24. (d)

29. (a) 231 nucleons (1MeV / nucl) = 231 MeV
   (b) 237 nucleons (1 MeV / nucl) = 237 MeV

32. (d)

36. $Q = (m_p - m_d - m_e)\, c^2$

   $Q = \{ [\, m_p - Zm_e] - [\, m_d - (Z + 1)\, m_e\,] - m_e \}\, c^2$

   $Q = (M_p - M_d)\, c^2$

44. The effects of the existance of virtual exchange particles can be predicted and these predictions can be confirmed experimentally.

49. $\Delta m_{\pi} = (274)\ m_e$

$\Delta E = (274) m_e c^2 = (274)(0.511\ \text{MeV}) = 140\ \text{MeV}$

$\Delta E = (140 \times 10^6\ \text{eV})(1.60 \times 10^{-19}\ \text{J/eV}) = 2.24 \times 10^{-12}\ \text{J}$

$\Delta t \geq h / 2\pi\Delta E$

$\Delta t \geq (6.6 \times 10^{-34}\ \text{J-s}) / [2\pi\ (2.24 \times 10^{-12}\ \text{J})] = 4.74 \times 10^{-23}\ \text{s}$

52. (a)

60. $Q = (m_p - m_e - m_d)c^2$

$Q = \{ [\, m_p - Zm_e] + m_e - [\, m_d - (Z - 1)\, m_e] \}\, c^2$

$Q = (M_p - M_d)\, c^2$

Sample Quiz

Multiple Choice. Choose the correct answer.

__ 1. A chain reaction occurs
A. in any uranium ore
B. when a critical mass is formed
C. on the Sun
D. under uncontrolled situations

__ 2. If the Q value is positive, the process is
A. endoergic
B. exoergic
C. energetic
D. none of the above

__ 3. All materials with atomic numbers greater than 92
A. are radioactive
B. created artifically
C. transuranium
D. all of the choices

__ 4. The fuel for nuclear fusion is
A. helium
B. hydrogen
C. uranium
D. plutonium

__ 5. This is the isotope which is responsible for most of the energy in a conventional reactor.
A. Pu-239
B. Pu-240
C. U-235
D. U-238

__ 6. This is the energy source on the Sun
A. aplha decay
B. beta decay
C. fission
D. fusion

__ 7. The "extra" energy in a beta decay goes to the
A. neutron
B. electron
C. proton
D. neutrino

Problems

8. Find the energy release in the following :

$$^{3}H + {}^{3}H \Rightarrow {}^{4}He + 2\,{}^{1}n$$

(3.016049 u) (4.002603 u) (1.008665 u)

9. Find the energy release when a ^{14}C atom undergoes a beta decay.

10. What is the Q value for the following :

$$^{14}N + {}^{4}He \Rightarrow p + {}^{17}O$$

(14.003074 u) (4.002603 u) (1.007825 u) (16.999131 u)

Answers to End of Chapter Quiz

Chapter 1

1. second ; meter ; kilogram
2. 3.25×10^5 m
3. m/s^2
4. D
5. D
6. D
8. (a) 17 mi/gal (b) 50 mi/h
9. 45 L ; 322 km
10. (a) 3.0 kg (b) 1.1×10^4 kg

Chapter 2

1. C 2. B 3. B 4. C 5. C 6. C 7. A
8. (a) 270 mi/h (b) 30 mi/h North
9. (a) 314 m (b) 12 m/s
10. (a) 22 m/s (b) 25 m

Chapter 3

1. A 2. C 3. D 4. C 5. C 6. B 7. C
8. -11.0 m **x** + 9.0 m **y**
9. (a) 4.0 s (b) 60 m (c) 42 m/s
10. (a) 3.3 s (b) 13 m (c) 40 m (d) 20 m/s (e) 12 m/s (f) $9.8\ m/s^2$

Chapter 4

1. C 2. B 3. D 4. C 5. C 6. A 7. D
8. $1.3\ m/s^2$
9. $3.3\ m/s^2$; 26 N
10. $0.89\ m/s^2$

Chapter 5

1. B 2. A 3. C 4. A 5. C 6. C 7. C
8. (a) 100 J (b) -39 J (c) 61 J (d) 61 J
9. (a) 3.4×10^4 J (b) 3.4×10^4 J (c) PE = 2.7×10^4 J ; K = 7.3×10^3 J
10. 7.5×10^3 W

Chapter 6

1. D 2. C 3. D 4. C 5. B 6. B 7. D

8. $v_1 = -1.32$ m/s ; $v_2 = 2.64$ m/s 9. 122 m/s

10. 1.3 m/s 72° above +x axis

Chapter 7

1. D 2. C 3. B 4. A 5. C 6. B 7. D

8. (a) 25 rad/s (b) 6.3 rad (c) $a_r = 3.1$ m/s^2 ; $a_t = 0.25$ m/s^2

9. (a) 2.2 m/s (b) 4.9 m/s 10. 2.0×10^{30} kg

Chapter 8

1. C 2. C 3. D 4. B 5. B 6. B 7. B

8. $F_1 = 2.2$ N ; $F_2 = 1.5$ N

9. (a) 19 rad/s^2 (b) 0.95 kg-m^2/s (c) 372 J (d) $K_m = 144$ J ; $K_M = 181$ J ; $K_p = 45$ J

10. (a) 9.2 m/s (b) 92 rad/s (c) 4.2 m/s^2 (d) would not roll and linear speed would be greater.

Chapter 9

1. C 2. D 3. C 4. B 5. B 6. A 7. D

8. 10.009 m 9. (a) 4.9×10^{-2} N (b) 0.338 N (c) 0.338 N

10. (a) 0.75 m/s (b) 4.22×10^3 Pa

Chapter 10

1. B 2. B 3. A 4. C 5. D 6. D 7. D

8. (a) 3.2 g (b) 2424 K 9. (a) 9.56×10^{-21} J (b) 462 K

10. 20.01 m

Chapter 11

1. A 2. C 3. C 4. C 5. D 6. D 7. D

8. 127 kcal 9. 10.9°C 10. 6.7×10^2 s

Chapter 12

1. B 2. A 3. B 4. D 5. C 6. B 7. C
8. (a) 1.5×10^3 J (b) 0 (c) 1.5×10^3 J (d) 146 K
9. -145 J/K 10. (a) 26% (b) 1.8 kW (c) 6.8 kW

Chapter 13

1. C 2. C 3. C 4. C 5. D 6. B 7. C
8. (a) 2.0 cm (b) 3.1 cm/s (c) 0.49 N/m (d) 4.0 s
9. 15 m 10. (a) 141 m/s (b) 141 Hz

Chapter 14

1. C 2. C 3. C 4. A 5. A 6. D 7. D
8. (a) 800 Hz (b) 842 Hz (c) 781 Hz
9. 115 dB 10. 4.53 m

Chapter 15

1. D 2. B 3. C 4. B 5. B 6. A 7. C
8. 1.62×10^5 N toward the -4.0 μC charge 9. (a) 0 (b) 1.1×10^5 V
10. (a) 2.2×10^{-10} F (b) 2.6×10^{-9} C (c) 1.6×10^{-8} J

Chapter 16

1. B 2. B 3. C 4. C 5. C 6. D 7. B
8. (a) 0.12 A (b) 4.5×10^{20} electrons
9. (a) 9.0×10^7 J (b) 21 A (c) 5.8 Ω 10. $230

Chapter 17

1. C 2. A 3. B 4. C 5. A 6. A 7. C
8. (a) 1.0 Ω (b) 8.0 W (c) 20 W 9. (a) 2.0 A (b) 0.67 A (c) 8.0 V
10. (a) 2.0×10^{-5} A (b) 7.4×10^{-6} A (c) 3.2×10^{-3} J

Chapter 18

1. A 2. D 3. B 4. D 5. A 6. D 7. B
8. (a) 1.0 m (b) 1.6×10^{-5} s 9. (a) 2.0×10^{-5} T (b) 6.0×10^{-4} N
10. 0.10 T into the page

Chapter 19

1. D 2. A 3. C 4. D 5. D 6. C 7. B
8. -2.7×10^{-5} V 9. (a) 0.40 V (b) 2.0 A (c) 0.80 W (d) 0.80 W (e) 0.16 N
10. (a) 9.0×10^{8} m (b) 4.3×10^{-8} m

Chapter 20

1. C 2. C 3. C 4. C 5. C 6. B 7. A
8. 45 V ; 3.2 A 9. 663 Ω
10. (a) 6.7 A (b) $V_R = 135$ V ; $V_L = 25$ V ; $V_C = 87$ V

Chapter 21

1. A 2. A 3. C 4. D 5. D 6. D 7. A
8. (a) 1.6×10^{8} m/s (b) 6.0×10^{14} Hz ; 267 nm (c) 15° 9. 62°
10. (a) 20° (b) 37° (c) 3.6 cm

Chapter 22

1. D 2. C 3. A 4. A 5. A 6. C 7. D
8. (b) 12 cm ; 6.0 cm 9. (b) 3.3 cm ; 1.3 cm 10. 86 cm

Chapter 23

1. B 2. B 3. A 4. B 5. A 6. C 7. C
8. 2.8 m 9. 0.38 m 10. 32°

Chapter 24

1. D 2. A 3. B 4. A 5. B 6. D 7. C
8. 6.7 cm 9. 1.6 cm 10. 27 cm

Chapter 25

1. B 2. A 3. B 4. C 5. D 6. B 7. B
8. (a) 1.8 m (b) 50 s 9. 0.91 c
10. (a) 3.6×10^{-23} kg-m/s (b) 5.3×10^{-14} J (c) 1.4×10^{-13} J

Chapter 26

1. D 2. B 3. B 4. D 5. D 6. B 7. D
8. 1.45×10^{-6} m ; 2.07×10^{14} Hz 9. (a) 1.5×10^{-6} m (b) 1.7 eV (c) 1.7 V
10. (a) 7.3×10^{14} Hz ; 4.2×10^{-7} m (b) yes

Chapter 27

1. B 2. D 3. D 4. A 5. B 6. B 7. B8.
$1s^2 2s^2 2p^3$ 9. 5.48 nm 10. 1.1×10^{-25} J

Chapter 28

1. B 2. A 3. A 4. C 5. B 6. D 7. D
8. (a) ${}^{145}_{61}Pm \Rightarrow {}^{4}_{2}He + {}^{141}_{59}Pr$ (b) ${}^{145}_{61}Pm \Rightarrow {}^{0}_{-1}e + {}^{145}_{62}Sm$
9. (a) 0.625 g (b) 3.54 g 10. 1518 MeV ; 7.95 MeV/nucleon

Chapter 29

1. B 2. B 3. D 4. B 5. C 6. D 7. D
8. 11.3 MeV 9. 0.16 MeV 10. -1.19 MeV